CLIMATE INFORMATION FOR PUBLIC HEALTH ACTION

Policy-makers are increasingly concerned about the impact of climate variability and change on the health of vulnerable populations. Variations and trends in climatic factors and extreme weather events impact many health outcomes, including malaria, heat stress and undernutrition.

Climate Information for Public Health Action is based on the premise that climate knowledge and information can help protect the public from climate-sensitive health risks. With a focus on infectious disease, hydro-meteorological disasters and nutrition, the book explores why, when and how data on the historical, current and future (from days to decades) climate can be incorporated into health decision-making. Created as a collaborative effort between climate and health experts, this book targets a broad technical public health community, alongside development practitioners and policy-makers engaged in climate change adaptation. It may also guide climate experts in the development of climate services tailored to health needs. Written in an accessible, informative style, while maintaining the highest technical and scientific standards, it will also be a valuable resource for students and academics studying and working in this emerging field of environment and health.

Madeleine C. Thomson is a Senior Research Scientist at the International Research Institute for Climate and Society, Earth Institute, and a Senior Scholar at the Mailman School of Public Health, Department of Environmental Health Sciences, at Columbia University, USA.

Simon J. Mason is Chief Climate Scientist at the International Research Institute for Climate and Society, Earth Institute, at Columbia University, USA.

Routledge Studies in Environment and Health

The study of the impact of environmental change on human health has rapidly gained momentum in recent years, and an increasing number of scholars are now turning their attention to this issue. Reflecting the development of this emerging body of work, the *Routledge Studies in Environment and Health* series is dedicated to supporting this growing area with cutting edge interdisciplinary research targeted at a global audience. The books in this series cover key issues such as climate change, urbanisation, waste management, water quality, environmental degradation and pollution, and examine the ways in which these factors impact human health from a social, economic and political perspective.

Comprising edited collections, co-authored volumes and single author monographs, this innovative series provides an invaluable resource for advanced undergraduate and postgraduate students, scholars, policy-makers and practitioners with an interest in this new and important field of study.

Ethics of Environmental Health
Edited by Friedo Zölzer and Gaston Meskens

Healthy Urban Environments
More-than-Human Theories
Cecily Maller

Climate Information for Public Health Action
Edited by Madeleine C. Thomson and Simon J. Mason

Environmental Health Risks
Ethical Aspects
Edited by Friedo Zölzer and Gaston Meskens

For more information about this series, please visit: https://www.routledge.com/Routledge-Studies-in-Environment-and-Health/book-series/RSEH

CLIMATE INFORMATION FOR PUBLIC HEALTH ACTION

Edited by Madeleine C. Thomson and Simon J. Mason

Routledge
Taylor & Francis Group
LONDON AND NEW YORK

earthscan
from Routledge

First published 2019
by Routledge
2 Park Square, Milton Park, Abingdon, Oxon OX14 4RN

and by Routledge
711 Third Avenue, New York, NY 10017

Routledge is an imprint of the Taylor & Francis Group, an informa business

British Library Cataloguing-in-Publication Data
A catalogue record for this book is available from the British Library

Library of Congress Cataloging-in-Publication Data
Names: Thomson, Madeleine C., editor. | Mason, Simon J., editor.
Title: Climate information for public health action / edited by Madeleine C. Thomson and Simon J. Mason.
Description: Abingdon, Oxon; New York, NY: Routledge, 2019. | Series: Routledge studies in environment and health | Includes bibliographical references.
Identifiers: LCCN 2018022507 | ISBN 9781138069633 (hardback) | ISBN 9781138069640 (paperback) | ISBN 9781315115603 (ebook)
Subjects: LCSH: Environmental health. | Public health--Environmental aspects. | Climatic changes--Health aspects.
Classification: LCC RA565.A3 C55 2019 | DDC 362.1--dc23
LC record available at https://lccn.loc.gov/2018022507

ISBN: 978-1-138-06963-3 (hbk)
ISBN: 978-1-138-06964-0 (pbk)
ISBN: 978-1-315-11560-3 (ebk)

Typeset in Bembo
by Sunrise Setting Ltd, Brixham, UK

CONTENTS

List of figures *xii*
List of tables *xv*
List of boxes *xvii*
List of case studies *xix*
List of contributors *xx*
Foreword *xxii*
Acknowledgements *xxiv*
List of abbreviations *xxvi*

1 Health priorities in a changing climate 1
Madeleine C. Thomson, Tamer Samah Rabie, Joy Shumake-Guillemot, John McDermott, Wilmot James and Chadia Wannous
1.1 Introduction 1
1.2 Policy drivers 2
1.3 Climate change and the global public health community 5
1.3.1 Climate impacts on infectious diseases 5
1.3.2 Climate impacts and disaster risk reduction 6
1.3.3 Climate impacts on nutrition 7
1.3.4 Impact of food systems on climate and health 7
1.3.5 From MDGs to SDGs 8
1.3.6 Universal Health Coverage 8
1.3.7 Climate shocks and conflict 9
1.3.8 The Global Health Security Agenda 9
1.3.9 Breaking down barriers to climate and health collaborations 11

1.4 Climate services 11
1.5 Conclusions 13

2 Climate impacts on disasters, infectious diseases and nutrition 16
Madeleine C. Thomson
Contributors: Delia Grace, Ruth DeFries, C. Jessica E. Metcalf, Hannah Nissan and Alessandra Giannini
2.1 Introduction 16
2.2 Hydro-meteorological disasters 21
2.3 Infectious diseases 24
2.4 Nutrition 27
2.5 Beyond the seasonal climate cycle to multiple timescales 35
2.6 Population vulnerability 37
2.7 Conclusions 37

3 Connecting climate information with health outcomes 42
Madeleine C. Thomson, C. Jessica E. Metcalf and Simon J. Mason
Contributors: Adrian M. Tompkins and Mary Hayden
3.1 Introduction 42
3.2 Climate information for use in health decision-making 43
3.3 Data issues 45
3.4 Exploring relationships 47
3.5 Linking climate to health outcomes 50
3.5.1 Statistical models 50
3.5.2 Mathematical models 53
3.6 Working with uncertain forecasts 55
3.7 Conclusions 56

4 Climate basics 59
Simon J. Mason
Contributors: Madeleine C. Thomson
4.1 Introduction 59
4.2 What is climate? 59
4.2.1 Temperature 62
4.2.2 Precipitation 66
4.2.3 Humidity 67
4.2.4 Wind 71
4.2.5 Solar radiation 74
4.2.6 Air quality 75
4.2.7 Other important meteorological parameters 76

4.2.7.1 Air pressure 76
4.2.7.2 Geopotential heights 76
4.2.7.3 Air chemistry 77
4.2.7.4 Sea, land and ice 77
4.2.8 Hurricanes, typhoons and other storms 78
4.2.8.1 Tropical cyclones 78
4.2.8.2 Extratropical cyclones 80
4.2.8.3 Tornadoes 82
4.3 How can climate be summarized? 82
4.3.1 If weather is what we get, what should we expect? 82
4.3.2 Aggregating weather data 83
4.3.3 How hot is hot? When does dry mean drought? 84
4.4 Conclusions 87

5 Climate variability and trends: drivers 89
Simon J. Mason
Contributors: Ángel G. Muñoz, Bradfield Lyon and Madeleine C. Thomson
5.1 Introduction 89
5.2 How does climate vary spatially? 89
5.2.1 Climate and altitude 90
5.2.1.1 Temperature and altitude 90
5.2.1.2 Humidity and altitude 95
5.2.1.3 Wind and altitude 95
5.2.1.4 Rainfall and altitude 96
5.2.2 Climate and latitude 96
5.2.2.1 Rainfall and latitude 96
5.2.2.2 Temperature and latitude 97
5.2.3 The effects of land and sea 99
5.2.3.1 Effects on temperature and the seasons 99
5.2.3.2 Effects on humidity and rainfall 100
5.2.4 The effects of land-surface type 101
5.2.4.1 Urban heat islands 101
5.2.4.2 Deforestation 102
5.2.5 Climate and spatial scale: How big is a heat wave or, a drought? 102
5.2.5.1 Spatial scales of temperature 103
5.2.5.2 Spatial scales of rainfall 104
5.3 How does climate vary temporally? 105
5.3.1 How does the time of day affect the climate? 105
5.3.1.1 Temperature 105

5.3.1.2 Rainfall 106
5.3.1.3 Winds 106
5.3.2 How long do weather patterns last? 107
5.3.3 What causes the seasons? 108
5.3.4 How do the seasons differ spatially? 109
5.3.5 How much does climate vary? 110
5.4 Why does climate vary temporally? 113
5.4.1 Internal causes of climate variability 113
5.4.1.1 Variability in earth's surface 114
5.4.2 External causes of climate variability 120
5.4.2.1 Volcanoes 121
5.4.2.2 Solar variability 121
5.4.2.3 Atmospheric composition 122
5.5 Conclusions 123

6 Climate data: the past and present 125
Simon J. Mason, Pietro Ceccato and Chris D. Hewitt
Contributors: Theodore L. Allen, Tufa Dinku, Andrew Kruczkiewicz, Asher B. Siebert, Michelle Stanton and Madeleine C. Thomson
6.1 Introduction 125
6.2 How are global weather and climate data produced and shared? 126
6.2.1 Global Observing System 126
6.2.2 Global Telecommunication System 128
6.2.3 Global Data Processing and Forecasting system 128
6.2.4 Global Atmospheric Watch 130
6.3 What types of meteorological data are available? 130
6.3.1 Direct measurements from climate stations 130
6.3.1.1 In situ station data 130
6.3.1.2 Gridded station data 131
6.3.1.3 Index datasets 132
6.3.2 Indirect measurements of climate by proxy, including by remote sensing 133
6.3.2.1 Historical proxy datasets 133
6.3.2.2 Satellite data 133
6.3.2.2.1 Satellite monitoring of rainfall 134
6.3.2.2.2 Satellite monitoring of temperature 138
6.3.2.3 Data from drones 138
6.3.3 Modelled data 138
6.4 What data and information are available? 140
6.4.1 Availability of historical and real-time data 140
6.4.2 Availability of historical and real-time information 142

6.4.2.1 Drought monitoring 144
6.4.2.2 Air chemistry and air quality monitoring 144
6.5 Conclusions 144

7 Weather forecasts: up to one week in advance 147
Simon J. Mason and Madeleine C. Thomson
Contributors: Heat Action Group, Kim Knowlton, Hannah Nissan, Ángel G. Muñoz, Carlos Perez Garcia-Pando and Jeffrey Shaman
7.1 Introduction 147
7.2 Why weather forecasts may be useful to the health community 148
7.3 Why is it so hard to predict the weather beyond a few days? 159
7.4 Given that it is hard, how do forecasters make predictions? 159
7.4.1 Observation 162
7.4.2 Analysis 162
7.4.3 Initialization 163
7.4.4 Integration 163
7.4.5 Post-processing 166
7.5 How accurate are weather forecasts? 166
7.5.1 Temperature 168
7.5.2 Rainfall 168
7.5.3 Tropical storms (cyclones, hurricanes and typhoons) 168
7.6 What weather forecasts are available? 170
7.6.1 Watches and warnings of hazardous and inhospitable conditions 170
7.6.2 Forecasts of unhealthy weather 171
7.6.3 Forecasts of suitable weather 171
7.7 Conclusions 172

8 Climate forecasts for early warning: up to six months in advance 175
Simon J. Mason
Contributors: Madeleine C. Thomson and Ángel G. Muñoz
8.1 Introduction 175
8.2 How do forecasters predict the next few months? 177
8.2.1 Why is the seasonal climate (sometimes) predictable? 178
8.2.1.1 The oceans 179
8.2.1.1.1 Tropical Pacific Ocean 179
8.2.1.1.2 Tropical Atlantic Ocean 180
8.2.1.1.3 Tropical Indian Ocean 180
8.2.1.1.4 Extratropical oceans 181
8.2.1.2 The land 181
8.2.1.3 Snow and ice 181

8.2.2 How are seasonal forecasts made? 181
8.2.2.1 Empirical prediction 182
8.2.2.2 Dynamical prediction 183
8.2.2.2.1 Observation 184
8.2.2.2.2 Analysis 185
8.2.2.2.3 Initialization 185
8.2.2.2.4 Integration 186
8.2.2.2.5 Post-processing 186
8.3 What seasonal forecasts are available? 189
8.3.1 Global Producing Centres of Long-Range Forecasts 189
8.3.2 Regional Climate Centres and Regional Climate Outlook Forums 191
8.3.3 National meteorological and hydrological services 192
8.3.4 Additional global products 192
8.4 Do seasonal forecasts work well? 193
8.5 Conclusions 195

9 Climate information for adaptation: from years to decades 199
Hannah Nissan, Madeleine C. Thomson, Simon J. Mason and Ángel G. Muñoz
Contributors: Glynn Vale, John W. Hargrove, Arthur M. Greene and Bradfield Lyon
9.1 Introduction 199
9.2 How increasing concentrations of CO_2 can impact health 200
9.2.1 Hydro-meteorological disasters 202
9.2.2 Infectious diseases 203
9.2.3 Nutrition 205
9.3 How climate-change projections are made 209
9.3.1 Downscaling 209
9.3.2 Multi-annual to multi-decadal prediction 211
9.4 How accurate are multi-annual to multi-decadal forecasts? 213
9.4.1 Climate model errors 213
9.4.2 How accurate are the predictions? 214
9.5 Conclusions 215

10 Climate information for public health action: challenges and opportunities 219
Madeleine C. Thomson and Simon J. Mason
Contributors: John del Corral, Andrew Kruczkiewicz, Gilma Mantilla and Cristina Li
10.1 Introduction 219
10.2 Climate services for health 220

10.3 Advances in technology 224
10.4 Institutional arrangements 228
10.5 Education and training 230
10.6 Conclusions 233

Index *235*

FIGURES

Source for all artwork is respective chapter author(s) except where stated throughout the book.

1.1 Direct and indirect interactions between climate and health – in part mediated by socio-economic effects 2
2.1 Basic, underlying and immediate causes of nutrition outcomes 28
2.2 Relationships between climate, agriculture, economy, nutrition and health in lower and middle-income countries (LMICs) 30
2.3 Rainfall variability in the Sahel at multiple timescales 36
3.1 Best forecast skill at multiple timescales with indications of the forecast ranges, timescales and spatial scales over which the forecasts are averaged 45
3.2 Understanding lags between climatic events and cases of disease 47
3.3 Relationship between annual malaria anomalies and December to January rainfall in Botswana 49
3.4 Schematic of the potential sources of uncertainty when using a weather/climate-sensitive disease model to simulate observed health outcomes 54
4.1 Latitudinal zones 60
4.2 Concentration of airborne bacteria in relation to humidity in Mali 68
4.3 Hourly temperature, dew-point temperature and relative humidity in Tucson, AZ 69

4.4 Locations of monsoon regions, as defined by areas that receive at least 70% of their annual rainfall during May–September (Southern Hemisphere [SH] winter / Northern Hemisphere [NH] summer) or November–March (SH summer / NH winter) 73
4.5 Global distribution of tropical cyclone tracks, 1991–2010 79
4.6 Typical structure of a mature extratropical cyclone 81
4.7 Frequency distributions of daily and monthly rainfall accumulations for Barbados for the wet season (August–November) 1981–2010 85
5.1 Average temperatures for January and July 1981–2010 90
5.2 Average rainfall for January and July 1981–2010 91
5.3 Temperature as a function of altitude over Brookhaven, NY, at 08h00 local time on 3 August 2017 92
5.4 Climatological (1981–2010) monthly average minimum temperature (°C) as a function of elevation for 18 stations in the Ethiopian Highlands 94
5.5 Rainfall (and snow) as a function of latitude 97
5.6 Temperature as a function of latitude 98
5.7 The annual range in temperature, calculated as the difference between the warmest and coldest mean monthly temperature 100
5.8 Example of a cold front, occurring on 26 November 2015 over part of the USA 103
5.9 Average daily-range in temperature for 1981–2010 106
5.10 Illustration of the causes of Earth's seasons 108
5.11 Year-to-year and day-to-day variability in temperature, as measured by the standard deviation, 1981–2010 111
5.12 The coefficient of variation of annual rainfall, 1981–2010 112
5.13 Sea-surface temperatures during a strong El Niño event (Dec 1997–Feb 1998) and a strong La Niña event (Dec 1998–Feb 1999) 115
5.14 Average December–February sea-surface temperatures, 1982–2017 116
6.1 Number of functioning meteorological stations by year in Rwanda that provide data in a) ENACTS and b) GPCC rainfall products 128
7.1 How well can we forecast severe heat or cold? 169
8.1 A measure of value of IRI's seasonal (three-month) average temperature and accumulated rainfall forecasts for 1997–2017 194
9.1 Impact pathways of rising CO_2 on social, ecological and health outcomes 201

9.2 Increase in the average minimum and maximum temperatures of each calendar month, from 1960 to 2016, at Rekomitjie Research Station in Zimbabwe 205
9.3 Sea-surface temperature anomalies and March–May (MAM) rainfall anomalies in Eastern Africa 208
10.1 Connecting information and people 228
10.2 Research findings can better inform decision-making if they take into account the different perspectives that influence decision-making 230

Access to full colour versions of these images can be found at: https://cipha.iri.columbia.edu/CIPHABOOK2019/Supplementary_Materials

TABLES

2.1 Health impacts of hydro-meteorological events 18
2.2 Climate sensitive infectious diseases 19
2.3 Stakeholder communities for different climate impacts on health pathways 21
2.4 Emerging infectious diseases that pose a significant risk to health security 27
3.1 Time horizons for decision-making in the health sector 44
3.2 Evidence for barriers to sharing of routinely collected public health data 46
4.1 Standardized Precipitation Index (SPI) thresholds and corresponding return periods (in years) for droughts of varying severity 87
5.1 Timescales of weather and climate variability and trends, their causes and sources of uncertainty 114
5.2 National de-trended (standardized) confirmed malaria cases (1982–2003) in Botswana during the malaria season (January–May) and their relationship to December–February rainfall 118
7.1 Definitions of meteorological forecasting ranges 150
7.2 Actions that could be taken in response to heat wave warnings at different timescales 153
7.3 Approximate average errors (in km) in predicting North Atlantic tropical storm tracks 170
8.1 Differences between weather and seasonal climate forecasts 176

8.2 Country-averaged December–February rainfall accumulations for Botswana 188
9.1 The main advantages and disadvantages of downscaling using statistical methods or regional climate models (RCMs) 210
9.2 Projections of near-term climate change (interannual-to-decadal, multi-decadal) with CMIP5 212
10.1 Summary of climate information of relevance to health sector decision-making 221
10.2 Resources in the climate change–climate and health arena 227

BOXES

1.1	Millennium Development Goals and Sustainable Development Goals	3
1.2	Threats to global health security	10
1.3	The Global Framework for Climate Services	12
2.1	Terminology for climate services	17
2.2	Climate and livestock	26
2.3	Climate and crops	31
2.4	Seasonality	32
3.1	Climate information opportunities	43
3.2	Use of systematic reviews	51
3.3	Five questions to consider when developing climate-driven models for decision-support	52
4.1	Weather, climate and climes	60
4.2	Measuring hot and cold	63
4.3	Which is the most useful humidity metric?	70
4.4	Ozone	75
5.1	El Niño – Southern Oscillation (ENSO)	115
5.2	Climate oscillations	120
6.1	Data-sharing policies	129
6.2	Data rescue	131
6.3	How do we measure ENSO?	132
6.4	Remote sensors	135
6.5	Drones	139
6.6	Flooding	141
7.1	Weather and climate forecasts	149

7.2 Sub-seasonal forecasts 151
7.3 Potential use of sub-seasonal forecasts in Heat Early Warning Systems 152
7.4 Measuring how good (or bad) forecasts are 155
7.5 Forecast formats 158
7.6 Ensembles 164
8.1 How does a seasonal climate forecast differ from a weather forecast? 176
8.2 Climate models 183
8.3 Tercile forecasts 187
9.1 A brief history of the science of global warming 201
9.2 Filtering the climate signal by timescales 206
9.3 Climate and health country profiles 212
10.1 Technology changes pre, during and post the MDG era 225

CASE STUDIES

2.1	Low birthweight	29
2.2	Drought in the Sahel	35
3.1	Plague, return of an old foe	48
3.2	Sources of uncertainty in modelling climate and malaria	53
4.1	Dispersal of pathogens and insect vectors – the importance of humidity and wind	68
4.2	The wind in the west: how the Onchocerciasis Control Programme followed the invasion of blackfly vectors from the Sahel to Sierra Leone	72
5.1	Elevation used in planning malaria control programmes	94
5.2	Impact of rainfall and the El Niño – Southern Oscillation on malaria in Botswana	118
6.1	Enhancing National Climate Services (ENACTS) data products	127
6.2	Seasonal malaria chemoprevention in the African Sahel	137
7.1	Dust storm impacts on health	150
7.2	Heat Action Plans and early warning systems help save lives in India	153
7.3	Weather forecasting techniques for flu forecasting	160
8.1	Understanding and predicting Latin Aedes-borne diseases in Latin America and the Caribbean using climate information	190
9.1	Tsetse – changes in climate in the Zambezi Valley: impact on Tsetse flies	203
9.2	The East African Paradox	207
10.1	Climate and health education in Colombia	231

CONTRIBUTORS

Editors

Madeleine C. Thomson leads the health work at the International Research Institute for Climate and Society (IRI), Earth Institute, Columbia University, New York, USA where she directs the IRI/World Health Organization (WHO) Collaborating Centre for Early Warning Systems for Malaria and Other Climate Sensitive Diseases. Her research is focused on the development of new data, methodologies and tools for improving climate-sensitive health interventions. She is also a Senior Scholar at the Mailman School of Public Health, and a visiting Professor at Lancaster University.

Simon J. Mason is Chief Climate Scientist at the IRI, Earth Institute, and a Senior Research Scientist of Columbia University, New York, USA. His work is focused on developing climate services at National Meteorological and Hydrological Services around the world. He was a lead author of the Global Framework for Climate Services Implementation Plan.

Chapter co-authors

Pietro Ceccato is a remote sensing expert with an interest in environmental drivers of health, based at the IRI, Columbia University, New York, USA.

Chris D. Hewitt is head of international climate service development at the Met Office Hadley Centre, UK.

Wilmot James is Visiting Professor of Pediatrics and International Affairs and Special Advisor: Global Health Security & Diplomacy, Columbia University, New York, USA.

John McDermott leads the Consultative Group on International Agricultural Research (CGIAR) Research Program on Agriculture for Nutrition and Health, International Food Policy Research Institute (IFPRI), Washington D.C, USA.

C. Jessica E. Metcalf is a demographer with broad interests in evolutionary ecology, infectious disease dynamics and public policy based at Princeton University, USA.

Ángel G. Muñoz is a climate scientist focused on the dynamics of climate predictability and the development of climate services, based at IRI, Columbia University, USA.

Hannah Nissan is a climate scientist focused on heat health and climate predictability from interannual to multi-decadal timescales at the IRI, Columbia University, USA.

Joy Shumake-Guillemot leads the WHO–World Meteorological Organization Climate and Health Office in Geneva, Switzerland.

Tamer Samah Rabie is Lead Health Specialist and the focal point for climate change and health at the World Bank, Washington DC, USA.

Chadia Wannous is coordinator of Towards A Safer World Network (TASW) for Pandemic Preparedness, Stockholm, Sweden.

FOREWORD

Climate change is identified as the most pressing concern for human health in the 21st century and climate change mitigation as the greatest public health opportunity of this generation. However, public health is not solely the purview of the health sector, a multi-sectoral approach is essential to reduce population vulnerability to climate change. This challenge calls for the engagement and commitment of a broad range of economic sectors – agriculture, urban planning, transport, energy, education and many more – to help societies build healthier lives for all. It is for this reason that the World Health Organization (WHO) and World Meteorological Organization (WMO) established a joint office for climate and health in 2014 to accelerate the use of climate science, services and information in the health sector.

The impacts of weather and climate events, such as droughts, floods, storms, heat waves, sea level rise and generalized warming, threaten the health of vulnerable populations around the world. Public health professionals, field epidemiologists, health management workers and health policy-makers have become increasingly interested in climate information that may help them better understand and manage the impacts of climate, which strongly affect their routine work. Opportunities are being sought to bridge climate intelligence with disease surveillance and control activities, the implementation and evaluation of public health interventions, and prevention measures for future risks. Climate experts and National Meteorological and Hydrological Services are keen to develop partnerships where their advancing science and technology can contribute to providing tailored tools and information to support the public health sector.

This publication promotes the multi-sectoral approach by bridging the knowledge gaps and creating a common understanding about the basic concepts – the data, methodologies and tools which are necessary to apply climate science to public health. It introduces why, when and how climate information can and should be incorporated into health research, policy and practice; examples of which are provided in the WHO-WMO Fundamentals and Case Studies of Climate Services for Health. The International Research Institute for Climate and Society at Columbia University is a long-standing partner of WMO in the development of climate services, and a WHO Collaborating Center for Malaria Early Warning and other Climate Sensitive Diseases. This publication demonstrates their unique capacity to support continued efforts by WHO and WMO to promote scientific approaches and impact-based services to managing climate risks for health.

Dr. Elena Manaenkova
World Meteorological Organization Deputy Secretary General

ACKNOWLEDGEMENTS

The book was a key community outcome from the Health and Climate Colloquium 2016[1] which was sponsored by the World Health Organization (WHO), including WHO – Special Programme for Research and Training in Tropical Diseases, the World Meteorological Organization, the Global Framework for Climate Services, the World Bank Group, the Nordic Development Fund, the International Development Research Centre, the CGIAR research programs on Climate Change, Agriculture and Food Security and Agriculture for Nutrition and Health, and the Earth Institute. The book chapters built off the presentations and panel discussion of the Colloquium. Many meeting participants have contributed to this book through co-authoring chapters, providing case studies and reviewing text. Chapter 1 was further elaborated by participants in the 2017 Consortium of Universities of Global Health 7 April 2017, Washington, D.C. panel discussion on Climate Change and the Health Benefits of the Sustainable Development Goals: Challenges & Opportunities. We thank Teddy Allen, Stephen Connor, Ruth DeFries, Tufa Dinku, Carlos Perez Garcia-Pando, Elisabeth Gawthrop, Richard Graham, Arthur Green, Ale Giannini, Delia Grace, John Hargrove, Jim Hansen, Mary Hayden, Kim Knowlton, Andrew Kruczkiewicz, Allie Lieber, Cristina Li, Bradfield Lyon, Gilma Mantilla, Sarah Molton, Ángel G. Muñoz, Jeffrey Shaman, Aisha Owusu, Michelle Stanton, Asher B. Siebert, Yohana Tesfamariam Tekeste, Rudi Thoemmes, Julian Thomson, Onny Thomson, Adrian M. Tompkins, Glyn Vale and Adugna Woyessa for their contributions and support for the project. A special mention goes to Ashley Bae, Jake Casselman, Dina Farone, Ximena Fonseca-Morales and Avalon Hoek Spaans, from the Masters in Climate and Society Program, Columbia University, for their assistance with the figures, the front cover, acronym list and review of the text.

We acknowledge with thanks the UK's Wellcome Trust for funding the writing of this book and for the provision of an open access online version. Additional funding was made available from the Columbia World Project, ACToday.

Note

1 https://iri.columbia.edu/healthclimate2016/

ABBREVIATIONS

AIDS – Acquired Immune Deficiency Syndrome
AMC – Ahmedabad Municipal Corporation
API – Application Programming Interface
AVHRR – Advanced Very High Resolution Radiometer
CDC – Center for Disease Control
CDR – Call Data Records
CFCs – Chlorofluorocarbons
COPD – Chronic Obstructive Pulmonary Disease
CMIP – Climate Model Intercomparison Project
CRD – Centre for Reviews and Dissemination
CRED – Centre for Research on the Epidemiology of Disasters
DHIS – District Health Information System
DMC – Disaster Monitoring Constellation
DNA – Deoxyribonucleic acid
DRM – Disaster Risk Management
DRR – Disaster Risk Reduction
ECMWF – European Centre for Medium Range Weather Forecasts
ECTS – European Credit Transfer Scheme
EM-DAT – Emergency Events Database
ENACTS – Enhancing National Climate Services
ENSO – El Niño-Southern Oscillation
EOD – Earth Observations Division
EPA – Environmental Protection Agency
ERSST – Extended Reconstructed Sea Surface Temperature
EWS – Early Warning System
GCOS – Global Climate Observing System
GCM – General Circulation Model

GCCHE – Global Consortium on Climate and Health Education
GDD – Growing Degree Days
GDP – Gross Domestic Product
GDPFS – Global Data Processing and Forecasting System
GFATM – Global Fund to Fight AIDS, TB and Malaria
GFCS – Global Framework for Climate Services
GFT – Google Flu Trends
GHG – Greenhouse Gas
GIS – Geographic Information System
GMHH – Global Monitoring and Human Health
GOES – Geostationary Operational Environmental Satellite
GOS – Global Observing System
GTS – Global Telecommunication System
HAP – Heat Action Plan
HEWS – Heat Early Warning System
HIV – Human Immunodeficiency Virus
ICDDR – International Center for Diarhoeal Disease Research
ICT – Information and Communication Techonology
IDE – Interactive Development Environment
IHR – International Health Regulations
IIPH-G – Indian Institute of Public Health-Gandhinagar
IMD – Indian Meteorological Department
IMS – Indian Meteorological Society
IPCC – Intergovernmental Panel on Climate Change
IRI – International Research Institute for Climate & Society
ITCZ – Inter-Tropical Convergence Zone
ITU – International Telecommunication Union
LBW – Low Birthweight
LMIC – Low and Middle-Income Countries
MAP – Malaria Atlas Project
MDG – Millennium Development Goals
MERIT – Meningitis Environmental Risk Information Technologies
MIT – Massachusetts Institute of Technology
MJO – Madden-Julian Oscillation
MME – Multi-Model Ensemble
MOOC – Massive Open Online Courses
MOS – Model Output Statistics
NAO – North Atlantic Oscillation
NASA – National Aeronautics and Space Administration
NCAR – National Center for Atmospheric Physics
NCD – Non-Communicable Disease
NDMA – National Disaster Management Authority
NHS – National Health Service (UK)
NICE – National Institute for Health and Clinical Excellence

NMHS – National Meteorological and Hydrological Service
NOAA – National Oceanic and Atmospheric Administration
NRDC – Natural Resources Defense Council
NWP – Numerical Weather Prediction
OA – Open Access
OCP – Onchocerciasis Control Programme
PDV – Pacific Decadal Variability
PEM – Protein-Energy Malnutrition
PHFI – Public Health Foundation of India
PMI – President's Malaria Initiative
ProMED – Program for Monitoring Emerging Diseases
RCM – Regional Climate Model
RCP – Representative Concentration Pathway
RCRC – Red Cross Red Crescent Climate Center
RSV – Respiratory Syncytial Virus
SDG – Sustainable Development Goals
SEIR – Susceptible, Exposed, Infected, Recovered
SIDS – Small Island Developing States
SMB – Shanghai Meterological Bureau
SOP – Standard Operating Procedures
SPI – Standardized Precipitation Index
SST – Sea-Surface Temperature
TB – Tuberculosis
UCL – Université Catholique de Louvain
UHC – Universal Health Coverage
UK – United Kingdom
UN – United Nations
UNESCO – United Nations Educational, Scientific and Cultural Organization
UNFCCC – United Nations Framework Convention on Climate Change
URL – Uniform Resource Locators
USAID – United States Agency for International Development
USGS – United States Geological Survey
WASP – Weighted Anomaly Standardized Precipitation
WBG – World Bank Group
WHO – World Health Organization
WIS – WMO Information System
WMO – World Meteorological Organization
WWW – World Wide Web

1

HEALTH PRIORITIES IN A CHANGING CLIMATE

Madeleine C. Thomson, Tamer Samah Rabie, Joy Shumake-Guillemot, John McDermott, Wilmot James and Chadia Wannous

1.1 Introduction

We live in an increasingly interconnected world. The rapidly increasing movement of people, pathogens, vectors, livestock, food, goods and capital across borders creates both economic opportunities and health risks.[1] Globalization is at the heart of this process, which is both driving, and subject to, accelerating global environmental and climate change. Climate interacts with health through a wide variety of direct and indirect mechanisms (Figure 1.1).

Floods, for example may directly lead to deaths through drowning, and can result in an increase in diarrhoeal and vector-borne diseases. They can also have a significant impact on food production, leading to increases in food prices and consequent reductions in household nutrition. Indirect social and economic impacts of climate and weather shocks may drive households into poverty, which is in itself a major determinant of poor health.[2] Epidemics, just like weather and climate, do not respect national borders and can threaten human health and social stability. Since the Millennium, the emergence of the coronavirus in 2003, the novel avian influenza (H1N1) in 2009, the Ebola virus in West Africa (2014–2016) and the Zika virus in the Americas (2015), amongst others, have demonstrated the speed at which infectious diseases can spread with devastating effect.[3] Emerging diseases are being joined by re-emerging threats such as plague and cholera. The presence of these old and new diseases is particular noticeable in rapidly urbanizing, underserved communities experiencing rising income inequalities, weakening health systems and significant social, ecological and climate changes.

Health is key to social and economic development, but as economies grow there is the potential to burn more fossil fuels, create more greenhouse gas (GHG) emissions and therefore exacerbate climate change. Thus, while, direct effects of climate on health are relatively easy to observe, indirect impacts of climate on society – and society on climate – are also important to consider.

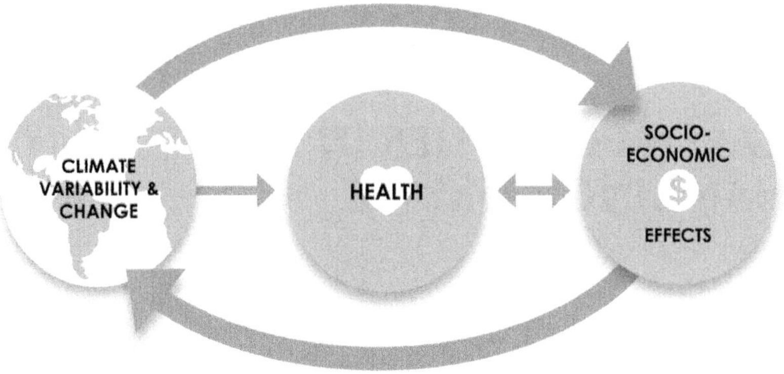

FIGURE 1.1 Direct and indirect interactions between climate and health – in part mediated by socio-economic effects

The international response to global health challenges has evolved over the last two decades to include this new understanding of a dynamic interaction between climate, environmental changes and health.

In this chapter we discuss some of the major policy drivers that have shaped, and continue to affect, international and national strategies where health and climate intersect. In particular we focus on the opportunities created for climate information and associated services to support public health decision-making in a changing climate. For example, historical information of the climate for a particular location is important for understanding geographic and seasonal risks whereas routinely updated monitoring products can be used to provide near-real time assessments. Weather and seasonal forecasts and longer-term predictions provide additional opportunities for understanding changing risks and enabling prevention strategies and/or early response.

Whilst not an exhaustive account, we provide context to the emerging climate and health dialogue which, to be successful, must tap into, challenge and if necessary modify ongoing policy processes.

1.2 Policy drivers

The establishment of the Millennium Development Goals (MDGs) in 2000 (Box 1.1) focused attention on the health needs of the poorest. This global commitment to poverty reduction resulted in increasing donor support for the prevention, control and elimination of infectious diseases with a focus on AIDS, TB and Malaria. For example, the President's Malaria Initiative (PMI), launched by President George W. Bush in 2005, provided an initial $1.2 billion of support over five years to 15 high malaria-burden countries in sub-Saharan Africa from the United States Agency for International Development (USAID). New public health

BOX 1.1 MILLENNIUM DEVELOPMENT GOALS AND SUSTAINABLE DEVELOPMENT GOALS

Millennium Development Goals (MDGs) were established by the global community following the Millennium Summit of the United Nations (UN) in 2000. The commitment was to combat poverty, hunger, disease, illiteracy, environmental degradation and discrimination against women. Each of the eight goals established had specific targets for delivery by 2015. Health was at the heart of the MDGs, dominating Goals, 1, 4, 5 and 6.

Goals

1. To eradicate extreme poverty and hunger
2. To achieve universal primary education
3. To promote gender equality and empower women
4. To reduce child mortality
5. To improve maternal health
6. To combat HIV/AIDS, malaria and other diseases
7. To ensure environmental sustainability
8. To develop a global partnership for development

The Sustainable Development Goals (SDGs) were established by the global community following the support of the UN General Assembly in 2014. Here the commitment is to end poverty, protect the planet and ensure prosperity for all. The 17 SDGs indicate a broader set of aspirations with a longer list of targets for delivery by 2030. In this agenda the significance of health is reduced while Goal 13 explicitly refers to climate change. Climate adaptation emerges as a cross cutting theme.

Goals

1. End poverty in all its forms everywhere
2. End hunger, achieve food security and improved nutrition and promote sustainable agriculture
3. Ensure healthy lives and promote well-being for all at all ages
4. Ensure inclusive and equitable quality education and promote lifelong learning opportunities for all
5. Achieve gender equality and empower all women and girls
6. Ensure availability and sustainable management of water and sanitation for all
7. Ensure access to affordable, reliable, sustainable and modern energy for all

8. Promote sustained, inclusive and sustainable economic growth, full and productive employment and decent work for all
9. Build resilient infrastructure, promote inclusive and sustainable industrialization and foster innovation
10. Reduce inequality within and among countries
11. Make cities and human settlements inclusive, safe, resilient and sustainable
12. Ensure sustainable consumption and production patterns
13. Take urgent action to combat climate change and its impacts
14. Conserve and sustainably use the oceans, seas and marine resources for sustainable development
15. Protect, restore and promote sustainable use of terrestrial ecosystems, sustainably manage forests, combat desertification, and halt and reverse land degradation and halt biodiversity loss
16. Promote peaceful and inclusive societies for sustainable development, provide access to justice for all and build effective, accountable and inclusive institutions at all levels
17. Strengthen the means of implementation and revitalize the global partnership for development

actors also emerged including the Global Fund to fight AIDS, TB and Malaria (GFATM). The GFATM was established as a funding mechanism to ensure the delivery of the Health MDGs, including malaria. Its problem-focused and transparent structure has given confidence to donors and country partners and thereby increased its impact.[4] Between 1998 and 2013 annual budgets for malaria control rose from \$100 million to \$2.7 billion.[5] The substantive resources made available by global donors and national governments for infectious disease control prompted a call for better surveillance systems to track impact. In 2008, the malaria coordinator of the GFATM suggested that the billions of dollars being used to fund malaria control were 'flying blind' and that investments in health surveillance systems, designed to inform disease prevention and control measures, would be cost-effective. He noted that surveillance would focus attention on the need to 'use data to assess the present situation, to target control measures and to evaluate their effectiveness'.[6] Health surveillance is now identified as an essential tool for health services delivery. The need for improved disease surveillance was also recognised in the creation of the International Health Regulations (IHR).[7] These regulations represent a binding international agreement involving all Member States of the World Health Organization (WHO). Their purpose is to prevent, protect against, control and provide a public health response to the international spread of disease in ways which avoid unnecessary interference with international traffic and trade.

1.3 Climate change and the global public health community

Preoccupied by the opportunity to make significant inroads into infectious disease prevention and control, the global health community was slow to respond to the rise of climate change on the political agenda.[8] However, building on the pioneering work of Tony McMichael and others,[9–12] in 2008 the World Health Assembly recognized climate change as one of the defining health challenges of the 21st century, and protecting health from its impacts as a priority for the public health community.

It is now well understood that the impacts of climate change on global health and development are manifold.[13] They result from changes in the extent and distribution of global warming (of the atmosphere and oceans); the changes in amount and variability of rainfall; the increased frequency and magnitude of extreme weather events and the extent and distribution of sea-level rise. Elevation of CO_2 in the atmosphere affects plant production, nutritional content and allergens as well as acidification of the oceans. In addition, society's response to climate change, in terms of both mitigation and adaptation, may of itself result in health impacts.

Climate change is substantially caused by increases in GHG emissions that are associated with the release of carbon from fossil fuel consumption. Rapid 'decarbonization' of society is key to climate change mitigation. The health benefits of reduced air pollution provide a powerful additional argument for the immediate economic and social benefits of reductions in carbon emissions at both the local and global level.[13] Consequently, health is now identified as a critical priority for protection from climate change as well as a co-benefit of mitigation in the United Nations Framework Convention on Climate Change (UNFCCC) Paris Climate Agreement of 2015. This accord committed countries to lowering GHG emissions in order to restrain warming below 2°C. However, in 2016 the United States decided to back away from the Paris Accord.[14] Despite this, economically powerful States, such as New York and California, continue to strengthen their climate change mitigation and adaptation activities. Paris reinforced the need for both developing and developed countries to create National Adaptation Plans (NAPs) to drive their adaptation agenda. Countries are currently developing NAPs; some of which include health as a priority.

1.3.1 Climate impacts on infectious diseases

Pressure to act on climate change is increasing in part because changes in the climate (and associated floods, droughts, heatwaves) are already being observed in many regions of the world.[15] These are associated with a range of health impacts. For example, an increased risk of respiratory, diarrheal, vector-borne and soft-tissue infectious diseases is observed amongst flood survivors and responders.[16] Many infectious diseases are climate-sensitive; climate acting as an important driver of spatial and seasonal patterns of infections, year-to-year variations in incidence (including epidemics), and longer-term shifts in populations at risk.[17] Increasingly those responsible for the development of disease control strategies have identified climate change as a challenge to their activities and climate information as a potential resource for programme planning.[18]

1.3.2 Climate impacts and disaster risk reduction

Epidemics and pandemics of air, water and vector-borne (i.e., transmitted by insects, ticks and snails) diseases may be identified as disasters in their own right; they have recently been included in the global institutional processes for disaster risk reduction (DRR).

A recent report[19] analysing trends in the past 20 years shows that 90% of disasters are weather-related with floods accounting for 47% of all hydro-meteorological disasters. These disasters claimed more than 600,000 lives, with an additional four billion or more people injured, left homeless or in need of emergency assistance. The report also noted that Africa is more affected by drought than any other continent.

Both climate and health shocks can have short term and long term (including intergenerational) impacts as evidenced from detailed household studies in Ethiopia.[20] The impacts of such disasters are experienced disproportionally by the poor[21] and mounting an effective response requires effective collaboration between health and DRR communities. The Sendai Framework for Disaster Risk Reduction 2015–2030[22] recognizes health as a key element of DRR and places emphasis on building resilient health systems through integration of all-hazards disaster risk management within health care and public health provision. This is a major advance on prior DRR frameworks which substantially ignored the role of the health sector in disaster response. Although civil defence agencies and non-governmental actors tend to dominate the disaster response community the health sector is needed to co-ordinate and promote health activities. This is most effective when it is undertaken with a people-centred approach. Whilst there is an urgent need to advance DRR initiatives in health, it is important to systematically integrate health care initiatives within DRR efforts to create a more comprehensive approach.

To advance the Sendai Framework for Disaster Risk Reduction and to bring the health and DRR communities together the Royal Thai Government, UN Office for Disaster Risk Reduction and the WHO organized an international conference on 10–11 March 2016, in Bangkok, Thailand. The conference report 'The Bangkok Principles'[i] offers opportunities for collaboration between all relevant sectors and stakeholders who wish to integrate health in DRR plans and strategies.[ii] Furthermore, 'The Bangkok Principles' fosters the inclusion of emergency and disaster risk management (DRM) programmes in health policies and strategies. Collaborative work between health and DRR communities now focuses on understanding disaster risk in all its dimensions of exposure, vulnerability and capacity. Joint risk assessment, profiling and prioritization as well as integrating health data into disaster loss databases is a vital part of this process, leading to comprehensive risk mitigation and reduction strategies.

Whilst climate change is moving up the health and DRR agendas the practical responses to climate shocks vary; reflecting the differences in community actors and institutional priorities. These differences are well illustrated if we consider the impacts of a major drought. Immediate practical responses to the drought may come from the agricultural or development community focused on household livelihoods, or through a national or local insurance scheme which provides resources to affected households based on agreed triggers such as those used for weather index insurance.[23]

1.3.3 Climate impacts on nutrition

Nutritious and sufficient food is the basis of health. Droughts can have a major impact on the nutritional status of vulnerable populations, particularly children, the elderly and women. These impacts, which may be the result of low calorie intake, insufficient micro-nutrients or infectious diseases such as diarrhoea, respiratory infections and parasites, have immediate and life-long effects.[24] Undernutrition persists in many countries but recently being overweight and/or obese has become an equally, if not more, important issue for health. This reflects an extraordinary transformation of global food systems especially in low and middle-income countries (LMICs). This transformation is largely determined by rising incomes, urbanization and greater economic activity in food systems in relation to processing, logistics and food retail. As a consequence, per capita consumption of meat, fish, vegetable, sugar and fats is increasing. This dietary transition has reduced calorie and micro-nutrient deficiency on the one hand, but dramatically increased obesity and associated non-communicable diseases (NCDs) on the other.[25] It is estimated that 50% of disease mortality and disability is now associated with NCDs such as cardiovascular disease and diabetes.[26] The impact of this transition is greatest in LMICs with enormous implications for health systems development. Health costs as a percentage of Gross Domestic Product (GDP) are typically 1 to 4% in low and lower-middle income countries. Pacific Island countries, with very high rates of obesity and NCDs, have health costs from 10–15% of GDP[27] mirroring what is observed in wealthier nations. The nutrition challenge is currently being addressed in the Decade for Action of Nutrition following the 2nd International Conference on Nutrition (ICN2).[28] Given the lack of past success in public campaigns to slow the increase in obesity and prevalence of NCDs, targets are usually very modest and often limited to a focus on children.[26,29]

1.3.4 Impact of food systems on climate and health

Crop production and supply chains that form the basis of food systems are impacted directly by climate. They are also major contributors to the GHG emissions that are the cause of human-induced (anthropogenic) climate change. Depending on the estimation method, from 20 to 30% of all GHG emissions come from food systems and this is even more significant in LMICs.[30,31] Thus, while food production, processing, transport and consumption must adapt to a changing climate, creating a low carbon food system that minimizes its exploitation of water and land is a critical part of any country's climate mitigation strategy. The nutrition community has been slow to engage with climate adaptation; the 2015 Global Nutrition Report[32] reported that, of those surveyed, only a small proportion of country nutrition plans made mention of climate related risks.

Likewise, the health sector has been surprisingly slow to get involved in nutrition and food issues in part because it requires broad engagement with the food industry. This needs to change given the rapidly increasing importance of private sector activities to both health and planetary sustainability.[33] Nutrition and food system issues will require a multi-sectoral approach with polices and investments recognizing the trade-offs of benefits and costs to different sectors (health, socio-economic, agri-food, environment). There is a need for policy levers that help focus the food industry on healthy diets and a safe and sustainable 'climate smart' food system for all.[34] Major investments in efficiency and effectiveness have the potential to reduce environmental costs, including climate related risks, while ensuring healthier foods (vegetables, legumes/pulses, fish, milk) at lower prices, especially for poor people and their children. Combined actions of both public and private sectors will be essential for progress in improving the sustainability and health of food systems.

1.3.5 From MDGs to SDGs

The Sustainable Development Goals (SDGs) (Box 1.1) provide a new paradigm in global health and development with their focus on new institutional relationships at the country level. This is particularly important for the health sector given that many of the drivers of health outcomes are the responsibilities of line ministries (e.g., transport, energy, agriculture), other institutions (including from the private sector) and communities that are not from within the formal health sector. From MDGs to SDGs there is increasing congruence between health, disaster risk reduction, nutrition and climate change policy processes.

1.3.6 Universal Health Coverage

Operationalization of the SDGs' health aspirations is being sought through implementation of programmes that support Universal Health Coverage (UHC). This initiative seeks to ensure that all people and communities can use effective, promotive, preventive, curative, rehabilitative and palliative health services that they need within manageable costs[35] although the detailed financing mechanisms needed remain to be established. Historically the health community has had little engagement and support from the major climate mitigation and adaptation funds. However, this is beginning to change, as illustrated by the work of the World Bank Group (WBG) whose task is to fight poverty worldwide through sustainable solutions. The WBG is now tackling the twin challenge of achieving UHC and reducing/managing climate related risks through the development of financing architectures that ensure joint development initiatives that serve both the health and climate agendas. Soon after the Paris Agreement, the WBG developed a climate action plan, which defines increased cross-sectoral support and reaffirms commitment for a one-third increase in climate financing under World Bank-funded projects; estimated at $16 billion by 2020. Taking this a step further, the Bank has also developed a specific action plan for climate and health, and has conceptualized ways of moving toward climate-smart lending in the health sector while supporting global efforts in achieving UHC. This push to increasing financing in areas related to climate change and health while adopting a UHC lens comes with the purpose of: establishing fair, efficient and sustainable health systems that are also adaptive; ensuring equity, affordability and quality of health services that are also resilient; and leveraging cross-sectoral climate-smart interventions to benefit health results and outcomes. This is particularly important in the case of lower-income countries which suffer the most from climate-related economic losses that at times may exceed 10% of their GDP.

UHC is sensitive to spatial and temporal variations in disease risk. For instance, Worrall and colleagues demonstrated that the cost of a malaria case prevented in a low malaria transmission year may be 20 times that of the cost of a case prevented in a high transmission year.[36] Improved targeting of health interventions is an emerging priority at a time of increasingly scarce resources. Furthermore, threats to health are increasingly understood as threats to societal well-being and security.

1.3.7 Climate shocks and conflict

The global food baskets, which provide cereal staples for the majority of the world's population are at risk from multiple shocks including climate shocks according to a recent study by Janetos et al. (2017).[iii] In their analysis, the authors consider the intelligence communities' failure to recognize the potential impacts of drought and high food prices in Syria immediately preceding the current civil war.[37] They note that while some analysts had been warning of the danger of large-scale migration, the broader community:

> *overlooked the links between infrastructure (the construction of a dam in Turkey to support agricultural self-sufficiency) and an extreme weather event in Syria (a drought that co-occurred with a global food price spike), with what has become a protracted civil war with extensive civilian casualties.*

At a more local level, the duration of drought has been shown to increase the likelihood of conflict for politically marginalized and agriculturally dependent groups – especially those residing in countries characterized by very low socioeconomic development.[38]

1.3.8 The Global Health Security Agenda

The emergence of a new global health security community initiated by the Global Health Security Agenda (GHSA) in 2014 focused first on pandemic health threats, but soon acknowledged that there is an inter-connected cascade of health security threat drivers. In addition to pandemic threats these include food insecurity, social unrest, biological, radiological, chemical and multi-hazard threats including cyber threats. This perspective requires a new, integrating lens which brings together the diverse communities that must respond at national and international level to such dangers. Using a whole-of-society approach the full range of threats is presented in Box 1.2. The GHSA offers to bring all of these challenges together at the highest level of government where a cross-sectoral approach to the health and well-being of citizens can be addressed. Accordingly, a high level of coordination across all eight community specialist domains is required to assure human survival and prosperity to meet the Sustainable Development and Global Health Security goals. The USA has been leading the GHSA but this will likely change as a result of funding cuts to the Centers for Disease Control.

Targeting health interventions requires a detailed understanding of the place-based nature of emergent health threats, their potential for rapid spread and the importance of early intervention, which can only be established with pertinent, high quality, information. Thus, achievement of SDGs in general, and the health targets in particular, is increasingly being associated with data-rich, evidence-based approaches. Climate data is identified as one source of necessary information for better management of climate sensitive health outcomes.[39]

BOX 1.2 THREATS TO GLOBAL HEALTH SECURITY

1. Biosecurity risks: prevention, preparedness and response to deliberate attacks on civilians with, or accidental release of, biological agents
2. Radiological and nuclear security risks: prevention, preparedness and consequence management of radiological and nuclear terrorism and large-scale radiological accidents
3. Chemical security risks: prevention, preparedness and response to large-scale acute chemical exposures of civilian populations, both intentional and accidental
4. Social cohesion risks: social rupture as a result of demographic change, extreme population mobility and competition for resources
5. Food security risks: an interconnected dimension, involving water, land, climate change, environmental sustainability and agricultural labour and use
6. Pandemic, infectious and zoonotic disease risks: prevention, preparedness and rapid emergency response to outbreaks such as Ebola and Zika
7. Multiple-hazard and general preparedness risks: multiple hazards or building infrastructure and capacity to respond to large-scale health threats
8. Cyber-security risks: accidental or deliberate attacks on health care and laboratory business systems that compromise patient record privacy and disrupt the functioning of medical technologies and emergency systems

1.3.9 Breaking down barriers to climate and health collaborations

Despite the fact that climate is an increasing concern for public health policymakers and national security advisors, few individuals and institutions are equipped to understand the impact of climate variability and change on health in a practical way or proactively manage the consequences of climate related impacts. In part, this gap in capacity is because of the desperate lack of applied research and training opportunities that could enable health decision-makers to understand the consequences of weather and climate on the outcomes they care about. Scale is a critical issue. Although climate change is a global phenomenon, the impacts and adaptation strategies needed are often quite local, requiring tailored information on both climate exposures and population health vulnerabilities.

Lack of access to, and effective use of, relevant climate information is central to the challenge. A gap analysis led by the International Research Institute for Climate and Society (IRI) in 2006 concluded that, despite climate's enormous impact on social and economic development in Africa (including health), climate information was not effectively incorporated into development decisions throughout most of the continent.[40] The problem identified was 'market atrophy': negligible demand coupled with inadequate supply of climate services which provide timely, tailored historical, current and future climate information and knowledge to decision-makers. The authors noted that climate data alone would not solve the problem. To be effective they must be created in response to user needs, i.e., engaging an empowered practice community that can create effective demand, and exist within an enabling policy environment were all essential to the development of a climate informed society.

This is beginning to change, driven by the multi-sectoral nature of the SDGs and the drive for policy integration across initiatives.

As an illustration of increasing policy integration, Roll Back Malaria's *Action and Investment to Defeat Malaria, 2016–2030*[5] highlights the linkages between climate and malaria in the SDGs. It stresses the need to build alliances between malaria programmes, ministries of health and relevant environmental and development partners (including national meteorological agencies) to secure access to climate adaptation funds. It also calls for national malaria programmes to integrate the management of climate-related risks into their programme activities. The document cites the example of Botswana which established an early warning system that integrates a seasonal rainfall forecast and climate/environmental monitoring information with population and health surveillance information.[41] The incorporation of climate information into the annual review and projection of malaria cases provided a four-month lead time for malaria response.

1.4 Climate services

Since the World Meteorological Organization World Climate Conference III of 2009, where the proponents of the Global Framework for Climate Services (Box 1.3) identified health as one of the priority climate sensitive sectors, there has

BOX 1.3 THE GLOBAL FRAMEWORK FOR CLIMATE SERVICES

The Global Framework for Climate Services (GFCS) was established to provide a credible, integrative and unique platform for guiding and supporting activities implemented within climate sensitive investment areas, notably agriculture, energy, disaster risk reduction, human health and water sectors *in support of both climate adaptation and mitigation*. The purpose of the GFCS is to provide a **framework for action** that enables and accelerates the coordinated, technically and scientifically sound implementation of measures to improve climate-related outcomes at national, regional and global levels.

The Goals of the Framework are:

1. Reducing the vulnerability of society to climate-related hazards through better provision of climate information
2. Advancing the key global development goals through better provision of climate information
3. Mainstreaming the use of climate information in decision-making
4. Strengthening the engagement of providers and users of climate services
5. Maximizing the utility of existing climate service infrastructure

The Framework's five foundational components, or pillars include:

- User Interface Platform: a structured means for users, climate researchers and climate information providers to interact at all levels.
- Climate Services Information System: the mechanism through which information about climate (past, present and future) will be routinely collected, stored and processed to generate products and services that inform often complex decision-making across a wide range of climate-sensitive activities and enterprises.
- Observations and Monitoring: to ensure that climate observations and other data necessary to meet the needs of end-users are collected, managed and disseminated and are supported by relevant metadata.
- Research, Modelling and Prediction: to foster research towards continually improving the scientific quality of climate information, providing an evidence base for the impacts of climate change and variability and for the cost-effectiveness of using climate information.
- Capacity Development: to address the particular capacity development requirements identified in the other pillars and, more broadly, the basic requirements for enabling any Framework-related activities to occur.

GFCS is a partnership with broad participation and reach; it serves as a voice for uniting many different parties, complementing the many existing programmes and initiatives contributing to climate services, building upon existing capacities and potentials, and providing momentum and tangible progress towards this fast-growing field. As such, it directly contributes towards the achievement of global and national goals identified in policy frameworks such as the Paris Agreement under the UNFCCC, the Sendai Framework for Disaster Risk Reduction 2015–2030, and the SDGs.

been a steady increase in awareness of the potential value of climate information and associated services to the health sector.[42]

At the heart of climate information is access to relevant high-quality data.[39] However, data is not enough; transforming data into information, understanding how to use the information and applying it appropriately for improved health outcomes are critical. The evolving policy context supports or detracts from this process. Greater congruence between development processes at the global, regional and national levels provides the opportunity to strengthen the use of data in evidence-based policy and practice. Improved access to climate knowledge and information will lead to a better understanding of the dynamic interaction between climate, environmental changes and health, the changing risks and new opportunities for the development of prevention and early response strategies. Climate change adaptation strategies for health are now being fostered by the Green Climate Fund (GCF), a new global fund created to support the efforts of developing countries to respond to the challenge of climate change. WHO is in the process of being accredited by the Fund so they can seek funding for health-related projects.

1.5 Conclusions

Here we propose the incorporation of climate information into routine epidemiological surveillance systems, early warning and risk assessment for climate-sensitive health outcomes as well as for all types of hydro-meteorological disasters, infectious disease emergencies and nutrition crises. To achieve this requires new and innovative mechanisms for strengthening observations, data management and sharing, development of relevant climate services, inter-sectoral collaboration, training and capacity building; all within an enabling policy environment. Our premise in writing this book is that improved management of health risks associated with climate variability (such as the heat early warning systems recently established in Europe and North America) increases adaptive capacity of the public health sector to longer-term climate change. Understanding the policy drivers that influence programmatic development, funding streams and new opportunities for inter-sectoral engagement can help ensure delivery of climate services that meet decision-maker needs.

Notes

i The Bangkok Principles for the Implementation of the Health Aspects of the Sendai Framework for Disaster Risk Reduction, www.preventionweb.net/files/47606_bangkokprinciplesfortheimplementati.pdf.

ii The Chair's Summary and the full report of the Bangkok conference 10–11 March 2016, www.unisdr.org/conferences/2016/health.

iii See www.bu.edu/pardee/files/2017/03/Multiple-Breadbasket-Failures-Pardee-Report.pdf.

References

1 Cilingiroglu, N. Health, globalization and developing countries. *Saudi Medical Journal* **26**, 191–200 (2005).

2 Marmot, M. Social determinants of health inequalities. *Lancet* **365**, 1099–1104, doi:10.1016/S0140-6736(05)71146-6 (2005).

3 Bloom, D. E., Black, S. & Rappuoli, R. PERSPECTIVE: emerging infectious diseases: A proactive approach. *PNAS* **114**, 4055–4059 (2017).

4 Clinton, C. & Sridhar, D. *Governing Global Health: Who Runs The World And Why?* (Oxford University Press, Oxford, 2017).

5 RBM. *Action and Investment to Defeat Malaria 2016–2030. For a Malaria-Free World.* 99pp (RBM, Geneva, 2015).

6 Grabowsky, M. The billion-dollar malaria moment. *Nature* **451**, 1051–1054, doi:10.1038/4511051a (2008).

7 WHO. *International Health Regulations* (third edition) (World Health Organization, Geneva, 2005).

8 St. Louis, M. E. & Hess, J. J. Climate change: impacts on and implications for global health. *American Journal of Preventative Medicine* **35**, 527–538 (2008).

9 McMichael, A., Haines, A., Slooff, R. & Kovats, S. *Climate Change and Human Health* (World Health Organization, Geneva, Switzerland, 1996).

10 McMichael, A. J. *et al. Climate Change 2001: Working Group II: Impacts, Adaptation and Vulnerability*. Chapter 9: Human Health 1–36 (IPCC, Geneva, Switzerland, 2001).

11 Patz, J. A., Gibbs, H. K., Foley, J. A., Rogers, J.V. & Smith, K. R. Climate change and global health: quantifying a growing ethical crisis. *EcoHealth* **4**, 397–405 (2007).

12 Kovats, R. S., Campbell-Lendrum, D. & Matthies, F. Climate change and human health: Estimating avoidable deaths and disease. *Risk Analysis* **25**, 1409–1418 (2005).

13 Costello, A. *et al.* Managing the health effects of climate change. *Lancet* **373**, 1693–1733 (2009).

14 Tollefson, J. NEWS: Trump pulls United States out of Paris climate agreement. *Nature* **546**, 198, (2017).

15 Herring, S. C. *et al. Explaining Extreme Events of 2016 from a Climate Perspective*. Special Issue: *Bulletin of the American Metereological Society* **99**, 1 (2018).

16 Ivers, L. C. & Ryan, E. T. Infectious diseases of severe weather-related and flood-related natural disasters. *Current Opinion in Infectious Diseases* **19**, 408–414 (2006).

17 Kelly-Hope, L. A. & Thomson, M. C. in *Seasonal Forecasts, Climatic Change, and Human Health* (eds. M.C. Thomson, R. Garcia-Herrera, & M. Beniston), 31–70 (Springer Science+Business Media, Berlin, Germany, 2008).

18 Thomson, M. C. *et al.* Using rainfall and temperature data in the evaluation of national malaria control programs in Africa. *American Journal of Tropical Medicine and Hygiene* **97**, 32–45, doi:10.4269/ajtmh.16-0696 (2017).

19 CRED/UNISDR. The human cost of weather related disasters: 1995–2015. 30pp (CRED/UNISDR, Brussels, 2015).

20 Dercon, S. & Hoddinott, J. in *Insurance Against Poverty* (ed. S. Dercon) (Oxford University Press, Oxford, 2005).

21 Field, C. B. *et al. Managing the Risks of Extreme Events and Disasters to Advance Climate Change Adaptation: Summary for Policy-Makers.* A Special Report of Working Groups I and II of the Intergovernmental Panel on Climate Change. 3–21 (IPCC, Geneva, Switzerland, 2012).
22 UN. *Sendai Framework for Disaster Risk Reduction 2015: 2030.* 36pp (2015).
23 Hellmuth, M. E., Osgood, D. E., Hess, U., Moorhead, A. & Bhojwani, H. in *Climate and Society Report* (ed. M. E. Hellmuth). 104pp (IRI, Palisades, New York, 2009).
24 Martins, V. J. B. *et al.* Long-lasting effects of undernutrition. *International Journal of Environmental and Public Health* **8**, 1817–1846 (2011).
25 Ng, M. *et al.* Global, regional, and national prevalence of overweight and obesity in children and adults during 1980–2013: a systematic analysis for the Global Burden of Disease Study 2013. *Lancet* **384**, 766–781 (2014).
26 IFPRI. *Global Nutrition Report 2016: From Promise to Impact: Ending Malnutrition by 2030* (IFPRI, Washington, DC, 2016).
27 WHO. Global Health Expenditure Database. Extracted by World Bank and accessed at http://data.worldbank.org/indicator/SH.XPD.TOTL.ZS on May 19, 2017 (2017).
28 FAO/WHO. *Rome Declaration on Nutrition.* Outcome document prepared for the Second International Conference on Nutrition (ICN, Rome, 2014).
29 Sacks, G., Shill, J., Snowdon, W., Swinburn, B., Armstrong, T., Irwin, R., Randby, S. & Xuereb, G. *Prioritizing Areas for Action in the Field of Population-based Prevention of Childhood Obesity* (World Health Organization, Geneva, Switzerland, 2012).
30 Vermeulen, S. J., Campbell, B. M. & Ingram, J. S. I. Climate change and food systems. *Annual Review of Environment and Resources* **37**, 195–222, doi:10.1146/annurev-environ-020411-130608 (2012).
31 FAO. *Climate Change Agriculture and Food Security* (FAO, Rome, 2017).
32 Thomson, M. C. Chapter 6. Climate and Nutrition (IFPRI, Washington, DC, 2015).
33 Whitmee, S. *et al.* Safeguarding human health in the Anthropocene epoch: report of The Rockefeller Foundation–Lancet Commission on planetary health. *Lancet* **386**, 1973–2028 (2015).
34 Gillespie, S. & van den Bold, M. Agriculture, food systems, and nutrition: meeting the challenge. *Global Challenges 1*, 1600002 (2017).
35 Tangcharoensathien, V., Mills, A. & Palu, T. Accelerating health equity: the key role of universal health coverage in the Sustainable Development Goals. *BMC Medicine* **13**, doi:10.1186/s12916-015-0342-3 (2015).
36 Worrall, E., Connor, S. J. & Thomson, M. C. Improving the cost-effectiveness of IRS with climate informed health surveillance systems. *Malaria Journal* **7**, 263, doi:10.1186/1475-2875-7-263 (2008).
37 Kelley, C.P., Mohtadi, S., Cane, M.A., Seager, R. & Kushnir, Y. Climate change in the Fertile Crescent and implications of the recent Syrian drought. *Proceedings of the National Academy of Sciences* **112** (11), 3241–3246 (2015).
38 von Uexkulla, N., Croicua, M., Fjeldea, H. & Buhaug, H. Civil conflict sensitivity to growing-season drought. *PNAS* **113**, 12391–12396, doi:10.1073/pnas.1607542113 (2016).
39 Thomson, M. C., Connor, S. J., Zebiak, S. E., Jancloes, M. & Mihretie, A. Africa needs climate data to fight disease. *Nature* **471**, 440–442 (2011).
40 IRI. *A Gap Analysis for the Implementation of the Global Climate Observing System Programme in Africa.* Report No. 06-01, 39pp (International Research Institute for Climate and Society, Palisades, New York, 2006).
41 Thomson, M. C. *et al.* Malaria early warnings based on seasonal climate forecasts from multi-model ensembles. *Nature* **439**, 576–579 (2006).
42 Guillemot, J. *Health Exemplar: Annex to the Implementation Plan for the Global Framework for Climate Services* (WMO, Geneva, 2014).

2

CLIMATE IMPACTS ON DISASTERS, INFECTIOUS DISEASES AND NUTRITION

Madeleine C. Thomson
Contributors: Delia Grace, Ruth DeFries, C. Jessica E. Metcalf, Hannah Nissan and Alessandra Giannini

> Whoever wishes to investigate medicine properly, should proceed thus: in the first place to consider the seasons of the year, and what effects each of them produces for they are not at all alike, but differ much from themselves in regard to their changes. Then the winds, the hot and the cold, especially such as are common to all countries, and then such as are peculiar to each locality.
>
> On Airs, Waters, and Places *by Hippocrates c. 400BC*

2.1 Introduction

The Zika virus epidemic that emerged in northeast Brazil in 2015 occurred during an unusually warm and dry year. Both natural climate variability as well as long-term trends were responsible for the extreme temperatures observed[1] and these climate conditions are likely to have contributed to the timing and scale of this devastating epidemic. Knowledge of this climate context is derived from analyses of large-scale global climate datasets and models, which provide policy-makers with broad insights into changes in hydro-meteorological extremes. However, societal response to epidemics works at multiple levels. For instance, policies and resource commitments may be developed at international and national levels, while targeted prevention and control efforts are managed at local levels by district health teams and community leaders. Adaptation to climate change also needs to be developed at multiple levels. National level information may be needed for planning, but an understanding of the local weather and climate that individuals and communities experience is also required. Once specific climate-sensitive health risks are identified, information on the past, present or future climate can be used to help mitigate risks and identify new opportunities for improved health outcomes. This information needs to be provided as a routine service if it is to support operational decision-making.

Climate services for health are an emerging technical field involving both the health and climate communities.[2] The Climate Services for Health Case Study Project showcases 40 studies that can help readers better understand what, how and why health-tailored climate services can support health solutions in managing climate risks. The publication emanating from the World Health Organization/World Meteorological Organization (WHO/WMO) Joint Office on Climate and Health presents a shared framework for developing climate services for health and highlights common needs and good practices.[i] A conceptual framework for the development of services is also emerging[3] along with relevant terminology (see Box 2.1).

However, to-date there is little concrete information on the value of such services and a significant gap remains between the potential for climate services to deliver actionable information routinely to health decision-makers and on the ground experience.

As our primary focus is human health it is helpful to consider *what is health*? In 1948, the WHO defined health in its Constitution with a phrase that is still used today: '*Health is a state of complete physical, mental and social well-being and not merely*

BOX 2.1 TERMINOLOGY FOR CLIMATE SERVICES[3]

- **Climate service coordinating bodies**, including the Global Framework for Climate Services, work to increase connections between climate information users and providers and to support the development of climate services in particular contexts.
- **Climate service users** employ climate information and knowledge for decision-making; they may or may not participate in developing the service itself. In some cases, climate information users may also pass information along to others, making them both users and providers.
- **Climate service providers** supply climate information and knowledge. Climate service providers may operate on international, national, regional or local levels and in a range of different sectors; they may be public or private, or some mixture of both.
- **Climate impact monitoring groups** meet to monitor and discuss evolving climate impacts and implications of forecasts for decision-making in particular contexts, especially with regard to health (e.g., Climate and Health Working Groups that monitor the incidence of climate-sensitive diseases) and food security. They generally include decision-makers, sectoral experts and representatives from practitioner communities.
- **Climate services** involve the direct provision of knowledge and information to specific decision-makers. They generally involve tools, products, websites or bulletins.

TABLE 2.1 Health impacts of hydro-meteorological events

	Health impact
Environmental	
Flood (general floods, flash floods and coastal floods)	The immediate health impacts of floods include drowning, injuries, hypothermia and animal bites.Vector-borne and water borne infectious diseases. Nutritional deficiencies through crop/livestock loss, livelihood disruption. In the medium-term, infected wounds, complications of injury, poisoning, respiratory problems from moulds, poor mental health, are indirect effects of flooding. Coastal floods may result in salt-water intrusion to drinking, hypertension, eclampsia.
Mass Movement (rockfall, landslide, avalanche, subsistence) often precipitated by high rainfall or warmer temperatures	The immediate health impacts of mass movement include loss of life, crush injuries.These may be followed by loss of livelihoods and mental health issues.
Meteorological	
Storm (hurricane, typhoon, cyclone) or local storm	The immediate health impacts of major storms include wind related crush injuries and loss of life. Further impacts may relate to flooding from associated rainfall (see above), infrastructure damage or sea surge.
Climatological	
Wildfire (forest fire, land fire)	Burns, respiratory morbidity, loss of life, cardiovascular, ophthalmic, loss of livelihoods, mental health problems. May increase mudslide risk.
Drought	Nutrition-related effects (including general malnutrition and mortality, micronutrient malnutrition, and anti-nutrient consumption (e.g., cassava)); water-related diarrhoeal diseases, airborne and dust-related disease (including meningococal meningitis); vector borne disease (including arboviral diseases transmitted by *Aedes aegypti* such as dengue and Zika); mental health effects.
Climate Extremes	
Heat wave	Heat stress, exhaustion, stroke.
Cold wave	Hypothermia, cardiovascular and respiratory health conditions, frost bite.
Extreme winter conditions	Snow, ice, frost, may result in increased car accidents, loss of life and injury.
Extreme humidity	High humidity may result in high levels of fungi and molds in the domestic environment which promote allergies including asthma. It may also result in fungal (e.g., aflatoxin) contaminated food stuffs with associated negative health consequences.

the absence of disease or infirmity'. More recently the Meikirch Model of Health posits that: '*Health is a state of wellbeing emergent from conducive interactions between individuals' potentials, life's demands, and social and environmental determinants*'.[4] The latter definition implies that there are multiple impact pathways for health.

In this chapter we introduce the three climate and health impact pathways that are central to our discussions throughout this book: i) the health outcomes of hydro-meteorological disasters such as floods, droughts and heatwaves (Table 2.1); ii) infectious diseases of humans and animals such as malaria, Zika, cholera and Rift Valley Fever (Table 2.2); iii) nutrition (Figure 2.1).

While there is clear overlap between all three pathways the stakeholder communities that are engaged in setting policy and responding in practice may differ substantially (Table 2.3).

TABLE 2.2 Climate sensitive infectious diseases

Trans-mission mechanism	*Type of pathogen*	*Disease*	*Pathogen*	*Epidemic potential*	*Transmission mechanism*	*Climatic or environmental transmission drivers*
Air-borne	Bacterial	Meningococcal meningitis	*Neisseria meningitides*	yes	Airborne aerosol	Aridity, dust, low relative humidity, temperature
Air-borne	Viral	Influenza	H1N1	yes	Airborne aerosol	Humidity, temperature
Food-borne	Bacterial	Gastroenteritis	*Salmonella* spp	no	Inappropriate food handling	Temperature
Vector-borne	Bacterial	Lyme Disease	*Borrelia burgdorferi*	no	Ticks *Ixodes* sp.	Rainfall, temperature, NDVI
Vector-borne	Filarial	Onchocerciasis / River Blindness	*Onchocerca volvulus*	no	Blackflies: *Simulium* sp.	Rainfall, temperature, NDVI, wind, river discharge
Vector-borne	Filarial	African Eye Worm	*Loa loa*	no	*Chrysops* sp. Forest canopy,	Forest soils, NDVI
Vector-borne	Parasitic	Malaria	*Plasmodium* sp.	yes	Mosquitoes *Anopheles* sp.	Rainfall, humidity, temperature, surface water puddles, river margins, irrigation, altitude, NDVI
Vector-borne	Parasitic	Schistosomiasis / Bilharzias	*Schistosoma* sp.	no	Snails e.g., *Bulinus Africanus*	Surface water, NDVI, temperature, rainfall, elevation

(*Continued*)

TABLE 2.2 (Continued)

Trans-mission mechanism	*Type of pathogen*	*Disease*	*Pathogen*	*Epidemic potential*	*Transmission mechanism*	*Climatic or environmental transmission drivers*
Vector-borne	Parasitic	African Trypanosomiasis / Sleeping Sickness, Ngana	*Trypanosoma brucei gambiensis*		Tsetse *Glossina* sp.	Gallery forests, savannah wood-land, temperature, NDVI
Vector-borne	Viral	Yellow Fever	*Flavivirus*	yes	Mosquitoes *Aedes, Haemagogus* and *Sabethes* sp.)	Rainfall, temperature
Vector-borne	Viral	Rift Valley Fever	*Phlebovirus*	yes	*Mosquitoes Aedes* and *Culex* sp.	Rainfall, humidity, surface water, temperature, NDVI
Vector-borne	Viral	Dengue and Dengue Hemorrhagic Fever	*Flavivirus*	yes	Mosquitoes *Aedes* sp.	Temperature, rainfall, humidity
Vector-borne	Viral	Zika	*Flavivirus*	yes	Mosquitoes *Aedes* sp.	Temperature, rainfall, humidity
Vector-borne	Viral	Chikungunya	*Flavivirus*	yes	Mosquitoes *Aedes* sp.	Temperature, rainfall, humidity
Water-borne	Bacterial	Cholera	*Vibrio cholerae*	yes	Faecal/oral route and filth flies e.g., *Musca* sp. via mechanical transmission	Water and air temperature, water depth, rainfall and conductivity, algal blooms, flooding, sunlight, SST
Water-borne	Viral	Gastroenteritis	*Rotavirus*	yes	Faecal/oral route and filth flies e.g., *Musca* sp. via mechanical transmission	Humidity, cool/ winter, dry months, low rainfall, water shortages, flood
	Bacterial	Trachoma	*Chlamydia trachomatis*	no	Flies e.g., *Musca sorbens* via mechanical transmission	Aridity, dust, low relative humidity, temperature

Abbreviations: NDVI, Normalized Difference Vegetation Index; SST, Sea Surface Temperature.

TABLE 2.3 Stakeholder communities for different climate impacts on health pathways

	Hydro-meteorological disasters of public health importance	*Infectious diseases epidemics/pandemics*	*Nutritional crisis*
Government	Multi-institutional disaster response (Office of the President)	Infectious disease departments or specific vertical programmes/Health Security Committee	Nutrition department, food security agency
National NGOs	National Society of the Red Cross Red Crescent	Local health NGOs	
International bilateral agencies	Infrastructure, engineering,	International partners responding to International Health Regulations e.g., WHO and CDC	World Food Programme/ FEWSNET/ SUN
Private sector	Insurance	Diagnostics	Food supplements
Academia	Engineering, public health/nursing/ emergency medicine	Public health, tropical medicine	Nutrition, food security
Specialist organizations	International Federation of the Red Cross Red Crescent Military	WHO, CDC, Military	World Food Programme, FAO

Abbreviations: CDC, Center for Disease Control; FAO, Food and Agriculture Organization; FEWSNET, Famine Early Warning System Network; NGOs, non-governmental organizations; SUN, Scaling up Nutrition; WHO, World Health Organization.

Other important pathways that are not significantly covered in this book include the direct health impacts of increases in atmospheric carbon dioxide on allergens[5] and air pollution. Nor do we cover the indirect impact of climate variability and change on economic growth and income inequalities[6] that affect the ability of governments to provide health services, and individuals to support a healthy lifestyle and, when necessary, to seek care.

2.2 Hydro-meteorological disasters

Hydro-meteorological disasters (droughts, floods, heatwaves, storms, etc.) are significant causes of mortality as well as of acute and chronic health issues.[7,8] The 1931 Central China flood disaster, which followed a period of extreme weather events, is the largest recorded disaster of the 20th century. The initial death toll was put at 150,000 from drowning; however, the total associated mortality is thought to have exceeded two million people, most of whom died from flood-related disease.[9] While much has been done in China to manage riverine flood disasters they remain a significant challenge for the population and government.[10]

The World Bank identified three investment areas for disaster mitigation: early warning, infrastructure and environmental buffering.[11] A common argument justifying investments in early warning is that early interventions are more cost-effective in reducing suffering and economic losses from disasters than late responses. It is therefore essential that early warning systems are tied to an effective early response. Forecast based Financing (FbF) is an innovative new approach to disaster risk reduction that seeks to scaffold disaster preparedness planning. Pre-allocated funding (necessary for rapid mobilization of pre-defined early action) is triggered to support 'just enough, just in time' preparedness, based on scientific (climate) forecasts. FbF is being promoted by the Red Cross Red Crescent (RCRC) Climate Centre to assist the mainstreaming of the early warning–early action model into RCRC disaster management worldwide.[12] Ability to demonstrate economic value from such early warning–early response systems is important to help ensure their long-term sustainability. Assessment of value requires the development of a counterfactual – an assessment of what would have happened if the forecast had not been available.

Measuring the impact of disasters on human and economic outcomes is important because it allows the humanitarian response to be based on concrete information on the type and scale of resources needed as well as to demonstrate the economic and social value of an early, organized response. Actions to measure the predicted or actual health impact of hydro-meteorological disasters pose a significant challenge (Table 2.1) and an agreed set of indicators are needed. According to the United Nations International Strategy for Disaster Reduction,[ii] primary indicators of disasters are: *deaths, missing, injured, exposed* and *economic loss,* with all five indicators taken as cumulative estimates without distinguishing between direct or indirect impacts. Secondary indicators provide greater refinements, for instance, identifying population movements and displacements such as *homeless, relocated* or *evacuated*, or other characterizations of the exposed population. By using these secondary indicators, the effects of disasters are counted in terms of the increased exposure of populations to additional morbidity and mortality, such as those derived from impacts on water and sanitation, vector-borne diseases, access to health care, depression, etc., as well as immediate lost lives and injury.

The impact of different types of disasters on health can be complex. Floods, for example, vary in their characteristics (see Box 6.6) and the vulnerabilities of affected populations may also differ. Areas at greatest risk of riverine flooding are low-lying flood plains or river beds located downstream from large catchment areas or dams. Areas at risk of flash floods include densely populated mountainous slopes such as those surrounding Freetown, Sierra Leone, where an estimated 1000 people were killed in a mudslide in August 2017 following an exceptional downpour in a deforested area. Coastal flooding is of greatest concern to countries like Bangladesh where a third of the country was underwater in the summer of 2017 following an unusually heavy monsoon. Many low-lying coastal areas globally are at risk of permanent flooding from sea-level rise in coming decades as a result of climate change.

People are often shocked to learn the extent of the toll on human life exacted by extreme heat each year. Heat waves were responsible for four of the ten deadliest natural disasters worldwide in 2015,[13] and remain the leading cause of declared weather-related disasters in Europe and the United States, outpacing hurricanes, floods and other dramatic weather-events that are usually considered more newsworthy.[14] However, definitions of heat waves and their impact vary from region to region depending on what temperatures the population normally experience; analysis of local health and weather data is necessary to understand temperature thresholds above which action should be taken. In practice, daily temperatures or apparent temperature (an index which describes the 'feels-like temperature' by also incorporating humidity [see Box 4.2[15]]) are most often used, although there are variations in approach.[14] For instance, temperature or apparent temperature are used for forecasts in the United States, Canada and in many European countries.

Hot nights have been associated with increased mortality in some of the most deadly heat waves, so thresholds for high night-time temperatures are often used as an indication of temperature mortality risk (for example in England, Montreal city and Poland). In some countries synoptic circulation systems associated with high heat-related mortality are used to supplement threshold-based heat forecast systems. To have a significant impact on mortality and health events hazardous hot conditions may be required to persist for two to three days to qualify for a heat wave. Once the characteristics of a heat wave at a particular location can be identified then there is the potential to create a locally relevant heat early warning system (see Case Study 7.2). Although heatwaves result in significant short-term health crises, outside of the tropics, seasonally cold weather kills 20 times as many people as hot weather.[16] Cold extremes are much less important in overall winter mortality than milder but non-optimum weather.

Drought disasters differ markedly from other natural hazards such as floods and heat waves – they are slow-onset events, which manifest over months or even years, over spatially diffuse areas, long before their many downstream impacts are felt.[8] Central to most definitions of drought is a deficit of water from a 'norm' for a given spatial area. However, the complexities of drought are reflected in its numerous definitions (over 150 according to researchers concerned with the issue).[17] These definitions differ according to the way drought is measured.

- *Meteorological drought* is defined based on the degree of dryness and the duration of the dry period due to less precipitation than normal.
- *Hydrological drought* is based on the impacts of precipitation shortages on surface or sub-surface (groundwater) water supplies.
- *Agricultural drought* links characteristics of meteorological or hydrological drought to agricultural impacts, where the amount of moisture in the soil no longer meets the needs of a particular crop.
- *Socioeconomic drought* occurs when the demand for a particular economic good exceeds supply as a result of weather-related shortfall in water supply and when water shortages begin to affect people.

The impact of drought on health outcomes may be wide-ranging, and involve multiple pathways including physical (e.g., dust inhalation) as well as nutritional and economic routes.

The disaster community treats epidemics of notifiable diseases (i.e., those reported to WHO) as disasters in their own right – whether or not the origin of the disaster is an unusual weather or climatic event. As a result, epidemics are included under 'natural hazards' in the Emergency Events Database (EM-DAT) that is run by the Centre for Research on the Epidemiology of Disasters (CRED) based at Université Catholique de Louvain (UCL) in Brussels, Belgium (www.emdat.be). Initiated with the support of WHO and the Belgian Government, EM-DAT has become a major global resource for the humanitarian community to rationalize decision-making for disaster preparedness. It also provides an objective base for vulnerability assessment and priority setting.

2.3 Infectious diseases

Epidemics of infectious diseases are those caused by other living organisms such as bacteria, viruses, worms, fungi or parasites living in or on human bodies. Many such organisms may normally be harmless, but under certain circumstances they become pathogenic and may cause mild to severe disease or even death.

Infectious diseases can be transmitted from person to person through a number of routes such as: i) in utero; ii) exchange of bodily fluids (e.g., during sex or blood transfusions); iii) via airborne aerosols (e.g., cough droplets); iv) via disease vectors (such as insects, ticks or snails); or v) through contaminated food or water. Climate may play a significant role in the transmission of many infectious diseases (Table 2.2). However, the importance of its role will depend on characteristics of the pathogen and its mode of dispersal.

In general, the rate at which infections may spread through a human (or animal) population can be captured by two quantities: the *basic reproductive ratio* (R_0), or number of secondary infections produced by a typical case of an infection in a population that is totally susceptible, and the *generation time*, or time between a case becoming infected, and causing other infections.[18] For dengue, R_0 has been estimated at around 5 while the generation time has been estimated at around two weeks. Using these values, if dengue was introduced into an urban population with little immunity, each infection would cause five new cases, so that in four weeks there would be 25 new cases, in six weeks, 125 – and rapid exponential growth will occur until immunity within the population restricts the pathogen's spread. Conversely, if R_0 were less than 1, the infection would go locally extinct: its continued occurrence is only possible from inward migration of infected individuals.

Both R_0 and generation time depend on innate characteristics of the pathogen itself and its mode of transmission, and human behaviour. For directly transmitted pathogens, the rate of contacts between individuals, as well as the probability of transmission on each contact, shape R_0; how long infected individuals tend to be infectious modulates both R_0 and the generation time. Climatic effects on these underlying features thus have the potential to modulate infection spread. Taking the dengue example above, if climatic conditions change to speed up mosquito life cycles, more frequent biting will increase the effective patterns of contact between individuals, thus increasing R_0, and that initial phase of exponential growth could be amplified.

Pathogens that cause epidemics in response to unusual climate conditions are characterized by high R_0 and short generation times, with vectors, pathogens and human hosts each responding quickly to changes in environmental characteristics. A single infective bite of a malaria transmitting mosquito may be sufficient to cause severe disease, even death, in a non-immune individual within two weeks. In epidemic malaria, disease and mortality statistics may be tightly correlated with changing environmental conditions. Rapid transmission cycles are the norm for bacterial, parasitic and viral infections. However, filarial worms, such as lymphatic filariasis (the cause of elephantiasis) or onchocerciasis (the cause of river blindness), have a much slower development rate in the human body and multiple infective bites are needed to infect a host sufficiently to cause disease. As a consequence, while climate drivers may significantly impact filarial worm transmission the relationship with human cases may be obscured by long and uncertain lags between infection and cases.

Climate and infectious disease analysis is dependent on accurate, long-term, historical disease data (either incidence or event data). Measuring the size and scale of infectious disease epidemics is challenging. A recent review identified over 1730 outbreaks/epidemics that have been reported in the WHO African region in the period 1970 to 2016.[19] There are, however, major inconsistencies in the epidemic records used in the report, which limit the use of this database for trend analysis. Challenges include: inconsistent definitions and thresholds for epidemics and outbreaks, inconsistencies between major epidemic databases and poor access to original epidemic reports. Data sources post-1980 indicate a marked improvement over earlier years, but changes in reporting and diagnosis continue to occur. Thus, extreme caution must be exercised when interpreting the trends in disease outbreaks and epidemics in Africa and in other data-poor regions. Painstaking efforts are needed to collate, manage and analyse historical disease data in relation to environmental data. Such analyses have been performed for meningococcal meningitis epidemics in Africa.[20] The Malaria Atlas Project (MAP, www.map.ox.ac.uk), initiated to support ongoing malaria control and elimination activities, has generated an extensive list of data sources that are publicly available and can be used in climate-malaria analyses. Historical records of infectious diseases in domesticated and wild animals are even more problematic than information on human cases. This makes assessment of zoonotic and emerging disease risk from changing patterns of climate and livestock farming in lower and middle-income countries particularly challenging (see Box 2.2).

BOX 2.2 CLIMATE AND LIVESTOCK

Delia Grace, International Livestock Research Institute
Nairobi, Kenya

Most of the world's farmed animals are kept by smallholders and pastoralists living in poverty in low and middle-income countries. Between 0.75–1 billion smallholders are concentrated in Asia and Africa.[21] Most are agro-pastoralists, integrating crops and livestock to harness ecological processes such as nutrient recycling and use of crop by-products. On the other hand, pastoralists rely mainly on livestock and use mobility to track scarce and shifting resources. Pastoralists are found from the drylands of Africa to the highlands of Latin America and the plains of Central Asia. They occupy around 25% of the global land area influencing ecosystems and contributing significantly to livestock products and economies.[22] Poverty, livestock keeping and infectious diseases are strongly and positively correlated. Around 60% of all human diseases and around 75% of emerging infectious diseases are zoonotic, that is, transmissible between humans and animals.[23] Impoverished livestock keepers and the consumers of the products they sell and produce bear a disproportionate burden of zoonosis and foodborne diseases.[24]

The high level of poverty and disease experienced by smallholders and pastoralists inevitably increases vulnerability to weather and climate through both direct and indirect effects. Direct effects include reduced livestock capacity to mount a response to infection (e.g., due to heat stress) as well as increased development rates of pathogens and vectors. Indirect effects, on the other hand, are associated with climate-driven ecosystem changes or socio-cultural and behavioural adaptations that could also amplify vector and pathogen development, or increase vector–pathogen–host contact. For example, drought-driven livestock movements have led to large increases of death from diseases to which the animals had no previous exposure.[25] Thirty-eight region-specific climate-sensitive diseases of high priority to poor people have been identified.[26] Among the most important diseases, food-and-waterborne zoonosis were prominent. Also notable were the parasitic endemic diseases that impose a high burden on productivity, water-transmitted leptospirosis and soil associated anthrax. Zoonoses play a prominent role in emerging infectious diseases and a number have been highlighted by the WHO as being amongst the most likely to cause severe outbreaks in the future (Table 2.4). Assessing, mapping and measuring climate-sensitive animal diseases is a pre-requisite to their better management using a 'One health' approach where veterinary, medical and public health professionals work together to prevent the spread of infection.[27]

TABLE 2.4 Emerging infectious diseases that pose a significant risk to health security[28] (many of these diseases are vector borne and/or zoonotic)

Disease	*Transmission*
Crimean Congo hemorrhagic fever virus	Ticks and livestock
Filo virus diseases (Ebola and Marburg)	Bats, person to person (respiratory)
Highly pathogenic emerging coronaviruses relevant to humans (Middle-East Respiratory Syndrome: MERS, coronaviruses and severe acute respiratory syndrome coronavirus)	Person to person
Lassa fever virus	Rats (urine, faeces) and person to person (blood/fluids)
Nipah virus	Bats to pigs to person (contaminated meat)
RiftValley Fever virus	Mosquitoes (*Aedes* and *Culex* spp.)
Chikungunya virus	Mosquitoes (*Aedes* spp.)
Severe fever with thrombocytopenia syndrome	Ticks and person to person (infected blood/fluids)
Zika virus	Mosquitoes (*Aedes* spp)

2.4 Nutrition

Good nutrition underpins good health. Maternal and infant nutrition depend on ready access to appropriate nutritious foods and the absence of diseases (such as those causing diarrhoea) that reduce the body's capacity to benefit from the food. Both dietary intake and disease are the result of broader drivers including household food security, the adequacy of care and feeding practices (e.g., breast feeding) as well as household environments and the quality of health services (Figure 2.1). Thus, nutritional status, which ranges across a spectrum from healthy to underweight/obese and nutrient deficient, is the result of a large number of basic, underlying and immediate causes, each of which may be climate sensitive.[29]

Consumption of nutritious foods (including all relevant micronutrients) is required to maintain the composition and function of an otherwise healthy individual within the normal weight range. At its most basic level, human health is sustained by vital vitamins and minerals that support normal cellular and molecular functions. Deficiencies in iron, iodine, folate, vitamin A and zinc are widespread and are common contributors to poor growth, intellectual impairments, perinatal complications and increased risk of morbidity and mortality,[31] particularly for women and children. Iron deficiency is the most common and widespread nutritional disorder in the world. As well as affecting a large number of children and women in developing countries, it is the only nutrient deficiency that is also significantly prevalent in developed countries. Over 30% of the world's population are anaemic, many due to iron deficiency in resource-poor areas. Anaemia is frequently

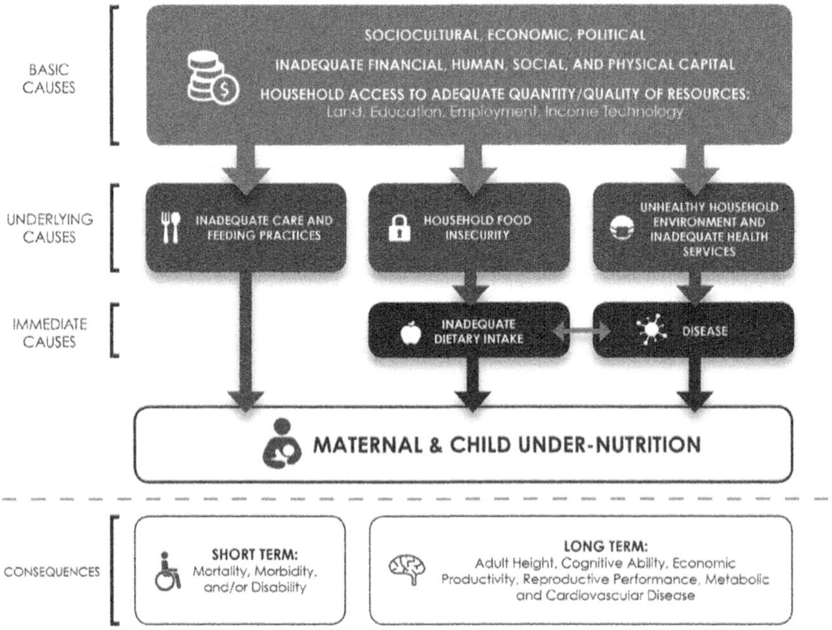

FIGURE 2.1 Basic, underlying and immediate causes of nutrition outcomes. Adapted from the United Nations Children's Fund (UNICEF)[30]

exacerbated by infectious diseases that interfere with iron absorption (e.g., malaria) or directly cause blood loss (e.g., hookworm).

Many factors influence the nutrient and calorific needs of individuals including age, gender, growth, disease states and genetic makeup. At any time, nutritional status may change rapidly as a result of loss in food consumption (e.g., due to loss in nutritious food availability or entitlement), an increase in nutritional requirements (e.g., due to exercise load or pregnancy) or a change in the body's capacity to absorb and metabolize necessary nutrients (e.g., infection with parasites, HIV, cancers, etc.).

Because of its complexity, nutritional status is measured using a range of clinical, social and anthropometric tools. For children, who are the most at risk of protein-energy malnutrition (PEM), anthropometric measures of weight and height have been used for many decades as indicators of malnutrition and potential mortality.[32] When nutritional disequilibrium occurs, it may be followed quickly (e.g., hours to days) by an alteration in processes that are associated with protein and energy metabolism (e.g., resulting in muscle fatigue) and subsequently manifests in changes over days/weeks in a bodyweight that is inappropriate for the child's height (wasting). Chronic malnutrition will result in bodyweights that are inappropriate for the child's age (stunting). 'Underweight' is also sometimes used as an indicator for malnutrition, but as a composite indicator of both stunting and wasting, it may be difficult to interpret. An important but often overlooked nutritional indicator is the weight of a child at birth. Low birthweight (LBW) is a strong indicator of infant morbidity and mortality as well as long-term health issues (see Case Study 2.1).

CASE STUDY 2.1 LOW BIRTHWEIGHT

Madeleine C. Thomson

Low birthweight (LBW) is defined by the WHO as a birthweight of less than 2500 g regardless of gestational age. The prevalence of LBW is 15.5% globally; 96 .5% of which is found in developing countries. Poor maternal nutrition (including iron deficiency) before and during pregnancy is a significant cause of LBW.[33] Other causes include risk factors, such as multiple births, smoking, indoor air pollution and parasitic infections, such as malaria. Measuring LBW is challenging in many settings as births may occur at home in the absence of trained midwives.

In areas of endemic transmission, malaria in pregnancy is associated with severe maternal anaemia and LBW babies. Malaria in pregnancy is commonly associated with placental infection which is highest in prima-gravidae (women who are pregnant for the first time); hence it is their babies who are at highest risk of LBW. The differential birthweight between prima and multi-gravidae can been used to assess the levels of malaria endemicity in a region.[34] In epidemic prone regions, where endemicity is low, both prima-gravidae and multi-gravidae have similar responses to infection and the birthweight differential is minimal, however in endemic regions the birthweight differential can be significant.

For example, a dramatic impact of the El Niño – Southern Oscillation (ENSO) on the LBW differential of prima-gravidae and multi-gravidae coming from an endemic region, Kagera can be observed in hospital records in Tanzania pre and post the 1997 El Niño. Data obtained from the delivery ward in Ndolage hospital, Kagera for the years 1990–1999 showed a significant increase in this differential (155 g) (P = 0.001) during April–August 1998, five months after the malaria epidemic which followed the exceptionally heavy short rains of 1997; the biggest ENSO event recorded at that time.[35]

The availability of nutritious foods is dependent on agricultural production, the nutritional content of crops and food safety; each of which may be affected by climate variability, trends and weather extremes. In the absence of irrigation, rainfall is needed to grow crops and even short drought periods at critical growth times can severely affect the harvest. Post-harvest processing may be significantly impacted by insects and pathogenic microbes (such as aflatoxins) that increase in warm and humid environments.[36] High temperatures may reduce the zinc and magnesium content of certain crops.[37]

The impact of climate on agriculture manifests in access to local, home-grown produce or changes in the price of food stuffs grown far away. Reduced production can ricochet throughout trade networks with severe consequences for prices and

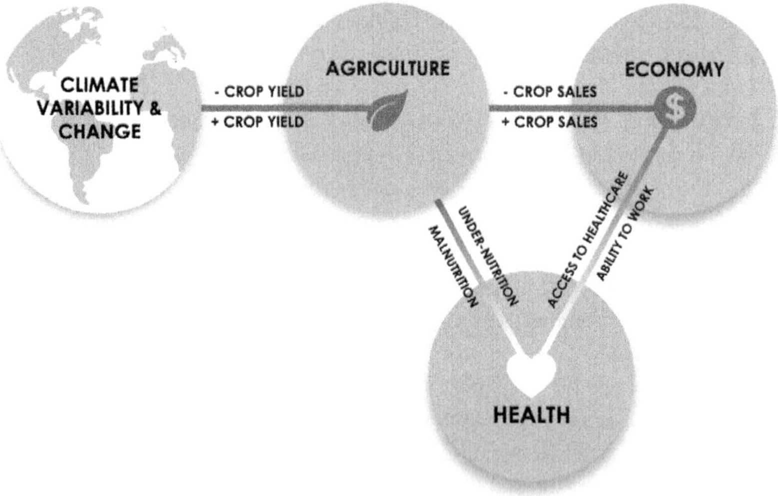

FIGURE 2.2 Relationships between climate, agriculture, economy, nutrition and health in lower and middle-income countries (LMICs)

availability. For example, droughts in Australia and Ukraine in 2007–2008 suppressed grain production and contributed to price spikes in many countries.[38] Significant price hikes for staple crops can have a major impact on nutritional status as families cut back on more nutritious foods to ensure that basic calorie needs are met.[39] Thus, in addition to direct effects, associated with food security and disease, the price of food and the capacity to purchase food, add significant new pathways to climate impacts on nutrition (Figure 2.2).

In response to these direct and indirect climate-related risks a renewed focus on traditional cereals is emerging, which could both achieve nutritional benefits and promote climate-resilient agriculture (see Box 2.3).

The nutritional value of a crop may also be impacted by pests and diseases that are weather sensitive. For example, many important food staples, e.g., maize and peanuts, are prone to contamination by mycotoxins. One example is aflatoxins, toxic secondary metabolites mainly produced by fungal pathogens *Aspergillus flavus* and *A. parasiticus*. Aflatoxins are known to cause liver cancer, and chronic exposure has been linked to other adverse health outcomes including growth faltering in children and kwashiorkor, a severe PEM disease.[44] The population potentially exposed to dietary aflatoxins is an estimated 4.5 billion people, predominantly those living in developing countries, with many chronically exposed at high levels.[45] Conditions that promote the production of the aflatoxin are hot weather and drought stress especially during the flowering and early grain-filling stages and heat and humidity during post-harvest storage. Other

BOX 2.3 CLIMATE AND CROPS

Ruth DeFries, Columbia University, New York, USA

Production of cereals, which globally provide almost half of the world's calories and cover more than half of all cropland area,[40,41] is vulnerable to drought, floods and other climate extremes. Cereals that are less sensitive to climate variability buffer against such shocks, as well as reduce vulnerability of small-scale farmers to fluctuations for home consumption.

Traditional cereals, such as millet and sorghum, which are grown and consumed in the sub-tropics, evolved in dry conditions with a C4 photosynthetic pathway that is more resilient to drought compared with C3 cereals such as rice and wheat (see below).[42] For instance, sorghum is highly nutritious in terms of iron, zinc and protein particularly when compared with rice.[43] C3 and C4 crops also respond differently to CO_2 fertilization that occurs in association with climate change, with possible consequences for nutritional content (see Chapter 9). Traditional crops such as sorghum and millet were substantially replaced by high-yielding rice and wheat following the Green Revolution beginning in the 1960s. Now, there is a renewed interest in traditional crops because of their nutritious value and climate resilience.

For example, farmers in the Central Highlands of India have recently demonstrated a renewed interest in traditional crops. The territory is characterized by a hot, sub-humid (dry) climate with a highly seasonal monsoon season. Fifty-four million people populate this agricultural region; nearly 70% of whom practice small-scale, rain-fed farming for subsistence and market. As in the rest of India, micro-nutrient deficiencies including iron and zinc are pervasive. The Central Highlands are highly vulnerable in terms of climatic variability and food security. Rice, the dominant monsoon crop in the region, is highly sensitive to climate variability and has the lowest content of protein and iron, when compared with traditional crops (e.g., sorghum and millet). A switch from rice to traditional coarse cereals could potentially provide more nutrition and improve climate resilience although many obstacles, such as production technologies, government subsidies, low yields, cooking habits and consumer preferences, complicate implementation of this win-win solution.[42]

- **C3 plants** are the most common (representing 95% of global plant biomass) and the most efficient at photosynthesis in cool, wet climates. They are most likely to benefit from increased growth due to CO_2 fertilization (see Chapter 9). Their growth becomes limited in hot and water stressed environments. C3 plants include key food staples such as rice, wheat, barley and soya bean.

- **C4 plants** are most efficient at photosynthesis in hot, sunny climates. They are less sensitive to CO_2 fertilization C4 plants are include maize, sorghum, sugarcane, tef and millet.
- **CAM plants** are adapted to avoid water loss during photosynthesis so they are best in deserts. A commercially grown example is agave. As they are adapted to hot and dry environments there is increasing interest in understanding how they can be better exploited for food and fibre in a warmer world.

weather-dependent food safety issues that may undermine nutritional status include *Salmonella* and *Campylobacter*.[46]

Floods may also impact on the underlying causes of poor nutrition by disrupting the availability and access of food to vulnerable groups, changing household activities (e.g., parents seek work away from home) resulting in reduced care for infants and children and changing the environment to increase exposure to disease. Floods may also limit access to health services. Major floods can significantly disrupt local and national economies impacting on household budgets and the tax revenues of governments used to support health systems.

Climate and other seasonal drivers (such as population movement, immune response, school holidays) may have an important effect on health outcomes in human populations. For infectious diseases, even small changes in seasonal drivers can drive large seasonal fluctuations in disease incidence, as a result of the amplifying effects of the inherent non-linear dynamics.[47] Considering the importance of such non-linear effects and the range of potential seasonal drivers, it is essential to distinguish the specific drivers of the seasonal pattern of disease if we are to understand the role of climate (see Box 2.4). The seasonality of the climate is described in detail in § 5.3.3 in Chapter 5.

BOX 2.4 SEASONALITY

C. Jessica E. Metcalf, Princeton University, USA

Seasonality drives all aspects of life in rural communities[48] including seasonal changes in nutritional status in populations that rely on household production and local markets. The 'hungry season' has long been identified as a period of severe stress to poor rural families that are unable to maintain body weight and function throughout the year. As nutrition is the foundation for health writ large, understanding how seasonal climate drives nutritional status is important for understanding many health issues.

A diversity of infectious pathogens, ranging from influenza to malaria, show clear seasonal fluctuations in incidence, with large numbers of cases concentrated at particular times of year.[49] Such seasonal patterns provide a uniquely repeatable probe for evaluating the association between climate drivers and health outcomes. Yet using this repeatable process to build evidence on climates role in driving disease is complicated by the diversity of ways by which health outcomes can be affected seasonally. For infectious diseases, the effects of seasonal fluctuations can range from direct effects of climatic conditions on pathogen transmission, indirect effects as a result of seasonal human biology or behaviour (including travel) and seasonal timing (with greater investment in control efforts) or disruption (e.g., floods or cyclones) of health system functioning.

Direct effects of climate variables on pathogen transmission

For directly transmitted infections (influenza,[50] chicken pox,[51] meningitis[52]) where transmission substantially relies on airborne movement between hosts (e.g., sneezing), seasonal fluctuations in incidence may emerge because humidity, or other climatic variables, shapes the way in which infectious particles fall out of the air. The onset of increased wintertime influenza-related mortality in the United States is associated with anomalously low absolute humidity levels during the prior weeks.[50] For food or water-borne pathogens, such as cholera or typhoid, seasonality of rainfall, which facilitates contamination, can shape seasonal incidence.[53] Cholera seasonality can also be formed by the biology of copepods whose association with the cholera bacteria is sensitive to seasonally fluctuating environmental variables.[54] Similarly, for *vector-transmitted infections* (malaria, dengue, Lyme disease), the biology of underlying insects or ticks, and their dependence on seasonal fluctuations, is key to understanding how climate seasonality modulates transmission[55]; the same is true of the seasonal biology of non-human reservoir species for some zoonotic pathogens (e.g., mice for hantavirus[56]). Finally, pathogenic species often interact, and although a focal pathogen might not, itself, be climate-sensitive, its abundance might depend on another species which is. For example, seasonally sensitive influenza increases the risk for invasive disease caused by *Streptococcus pneumoniae*,[57] but this latter pathogen might not, itself, be affected by seasonal drivers.

Indirect effects via human biology or behaviour

Many aspects of human biology relevant to health status are seasonal. Seasonality in immune function (e.g., associated with vitamin D metabolism and sunlight[58]) is perhaps the most obvious driver of seasonal fluctuations in human health status. In the 'meningitis belt' in Africa, seasonal dust storms which occur during the protracted dry season have been suggested as an important driver of meningitis outbreaks. The proposed mechanism which underpins this relationship is that damage to the human pharyngeal mucosa from the dry and dusty weather eases bacterial invasion[52] (see Chapter 7). The 'hungry season' may also affect susceptibility to infection. Poor nutrition prior to the harvest season has been suggested as a possible driver of respiratory syncytial virus (RSV) seasonality in the Philippines, for example.[59] Less directly, conception is seasonal all around the world, for reasons which remain poorly characterized. Since the resulting seasonality in births will result in a seasonality in the replenishment of individuals with no immunity to infection, birth seasonality could allow greater spread of immunizing infections at particular times of the year.[60] A strong relationship between influenza prevalence in the month of birth and prematurity in part accounts for the seasonality of the length of gestation: infants conceived in the USA in May have the shortest gestation, they are likely to be due in mid-February, which is the height of the flu season.[61] Pre-term neonates are also likely to be of low birthweight. Thus seasonality in births interacts with seasonality in infection risk to shape the burden of disease.

Human behaviour is also seasonal in ways that can shape exposure to infectious diseases. A classic example is schooling, known to be a key driver of transmission of directly transmitted childhood infections like measles, as transmission is magnified when children aggregate in schools during term times.[62] Seasonal migration linked to agriculture, fisheries and pastoralism[63] is also widespread, and may shape measles[64] and meningitis[65] dynamics in sub-Saharan Africa. Travel associated with seasonal holidays has also been found to impact the speed at which pathogens are introduced to new communities.[66]

Health system functioning

One of the largest footprints on many infectious diseases' incidence globally is the impact of control efforts. Since control efforts tend to focus on time periods in which transmission is most intense (e.g., indoor residual spraying may be concentrated during the season of greatest mosquito abundance) timing of interventions is an important consideration in evaluating seasonality in infectious disease incidence. Conversely, events such as hurricanes or flooding may reduce the functionality of health systems during particular times of the year, with roads impassable and health care delivery intractable. The timing of vaccination weeks in Madagascar illustrates this well. Here the timing of interventions, when mothers and children receive many important health care components, is set to be either side of the hurricane season. The timing of vaccination may modulate seasonality in health outcomes.[67]

Observations of seasonal effects on health status have revealed astonishing impacts and intriguing new drivers of human morbidity and mortality. For example, rural Gambian children born during the rainy season are up to ten times more likely to die prematurely in young adulthood than those born in the dry season.[68] Nutrition-related epigenetic regulation in the early embryo may be a highly plausible mechanism for this seasonality in mortality.[69] Epigenetic processes describe changes to the genome that can alter gene expression without changing the underlying DNA sequence and there is strong evidence that these changes can be influenced by a diverse array of intrinsic and environmental factors, including age, disease, stress, exposure to pollutants and nutrition.

Given the importance of seasonality in the climate and other factors in driving disease any study of climate and health interactions should start with an exploration of seasonal drivers.

2.5 Beyond the seasonal climate cycle to multiple timescales

Beyond the cyclical seasonal structure weather and climate vary on multiple timescales, from specific weather events (minutes to hours) to daily, seasonal, decadal and long-term climate change timescales. The best studied example of the way the climate interacts with society on multiple timescales is the African Sahel (Case Study 2.2). These different timescales will be discussed in greater detail in Chapters 5–9.

CASE STUDY 2.2 DROUGHT IN THE SAHEL

Alessandra Giannini, IRI, Columbia University, New York, USA

The majority of the rains in the African Sahel fall between July and September. The amounts vary considerably from year to year, but are potentially predictable given the proven influence of global sea surface temperatures, including those in the Central and Eastern tropical Pacific associated with ENSO events.

Extreme drought years occurred in 1972 and 1982–1984 (Figure 2.3). These droughts, which were embedded in a longer drought cycle, resulted in widespread food insecurity and severe malnutrition, population movement and loss of traditional livelihoods as the herds of pastoralists were decimated. Human mortality due to famine was rife.[70] The longer (decadal) drought cycle is also associated with sea surface temperatures, only this time in the oceans around Africa, including the warming Indian Ocean.[71] The observable long-term drying trend for the region might suggest that the climate change signal for the region is towards reduced rainfall.

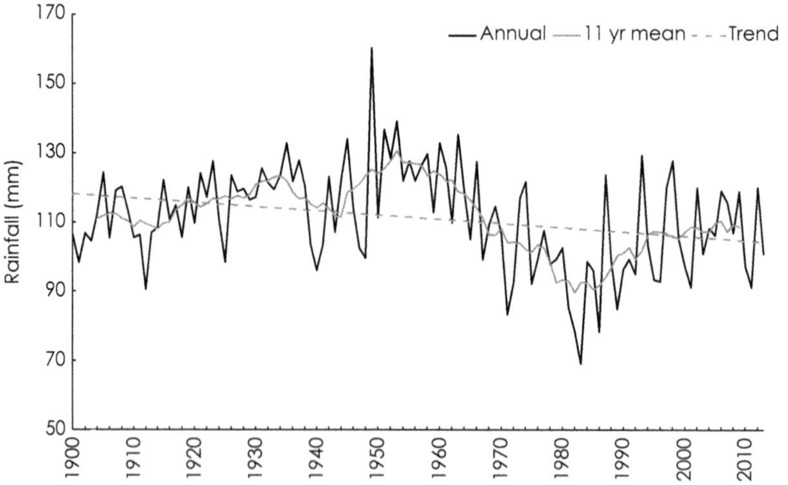

FIGURE 2.3 Rainfall variability in the Sahel at multiple timescales. Rainfall data from UEA/.CRU/.TS3p23

However, there is now compelling evidence that the 1970s–1980s' drought was in part due to the impact of aerosols from industrialized societies on the North Atlantic Ocean sea-surface temperatures.[72] These multiple timescale influences on seasonal climate have significant implications for climate change attribution. The recent recovery of rainfall in the Sahel, after the devastating droughts, may be in part a natural decadal cycle with origin in internal oceanic processes, in part attributable to the reduction in aerosols due to pollution control. One outcome of the drought was a greater than 80% decline in malaria prevalence in the semi-arid areas of northern Senegal and Niger from the early 1960s to the mid 1990s.[73] The drought, and associated land use changes, resulted in a loss of vector breeding sites, lower vector survivorship and ultimately the disappearance of the malaria vector species *Anopheles funestus*. The overall result was a shortening or reduction in intensity of the malaria transmission season.

The idea that modification of sea surface temperatures (SSTs) by anthropogenic emissions is the driving force behind late 20th century drought contrasts dramatically with earlier perceptions. Prior to the focused climate research, responsibility for these regional drought disasters had been put predominantly upon the shoulders of local peasant farmers, whose overuse of the land was deemed to have resulted in reduced vegetation cover, the advance of the desert and localized impacts on the regional climate regime.[74] Emerging evidence relating drought to anthropogenic aerosols shifts the blame to industrialized countries while indicating that the Sahel's future might be wetter under climate change in the absence of aerosols.

Different external drivers are associated with these timescales – weather, for instance is predominantly associated with atmospheric influences whereas seasonal and decadal changes in the climate are driven primarily by changes in the temperatures of the oceans. Long-term trends in climate, consistent with climate change impacts, are strongly influenced by anthropogenic forcing, including carbon emissions. Predictability at different timescales varies according to these underlying drivers (see Table 5.1).

2.6 Population vulnerability

Some groups of people are inherently more vulnerable to the impacts of weather and climate events and associated environmental hazards than others. The elderly or very young, the sick, and the physically or mentally challenged are vulnerable. A number of physiological, psychological, cultural and socioeconomic factors contribute to this vulnerability including poverty and social marginalization, Women, who globally are socially and economically disadvantaged, may be vulnerable to weather and climate extremes through reproductive factors (pregnancy and lactation) as well as sexual and domestic violence, which commonly follow disasters.[75] Their greater responsibilities as care-givers to other vulnerable groups (elderly parents, children and the sick) also increases women's vulnerability to disasters and in part explain their higher risk of dying.[76] Older adults have a higher prevalence of certain diseases, medical conditions and functional limitations that put them at risk of hydro-meteorological events; these include increased social isolation, poverty and higher sensitivity to extreme heat.[77] The effect of hot weather on the human body is determined not only by temperature, but by humidity, wind speed, cloud cover and night-time vs day-time conditions. Heat exhaustion may be followed by heat stroke when temperatures are extreme. Hot and humid nights are particularly associated with increased mortality as individuals' ability to stay cool may be limited. Children are also particularly vulnerable to excessive heat because of their small size and dependency on others.

2.7 Conclusions

Natural climate variability has always been important in human development. While year-to-year variations in rainfall and temperature cause significant challenges to many aspects of human health and well-being the stark seasonality of the climate in rural areas in many developing countries is the primary source of climate impacts on health. Here populations undergo seasonal (i.e., highly predictable) dramatic changes to health and well-being including hunger, nutritional deficiencies, disease, livelihood loss, migration and debt. Seasonal forecasts may indicate likely shifts in the probable outcome of the rainy season but the underlying season will still dominate the health response. Climate change may impact on the length and intensity of the rains, but the underlying seasonal patterns will remain as these are determined by factors that are not amenable to significant change. Even in urban areas in developed countries seasonality governs many aspects of health, from the timing of epidemic flu and heat-associated cardiovascular crises to the risk of fractures from falls on ice. The occurrence of extreme events, such as those precipitated by hurricanes or cyclones, has a marked seasonality to their occurrence. Hippocrates' statement, over 2000 years ago, that medical students should understand the importance of seasonality when considering health issues[78] is still relevant today. In the context of a changing climate seasonal challenges will continue to be significant while new threats to health emerge.

In Chapter 2 we have explored some of the many ways in which climate impacts on health; focusing on the health outcomes of hydro-meteorological disasters, infectious diseases and nutrition. Having identified a problem that is climate-sensitive, the subsequent chapters in this book should help the reader to consider how, when, where and why climate information might be used to mitigate some of the risks and improve the health of vulnerable populations.

Notes

i https://public.wmo.int/en/resources/library/climate-services-health-case-studies.
ii www.irdrinternational.org/wp-content/uploads/2015/03/DATA-Project-Report-No.-2-WEB-7MB.pdf.

References

1 Muñoz, A. G., Thomson, M. C., Goddard, L. & Aldighieri, S. Analyzing climate variations at multiple timescales can guide Zika virus response measures. *GigaScience* **20165**, 41, doi:10.1186/s13742-016-0146-1 (2016).
2 Guillemot, J. *Health Exemplar: Annex to the Implementation Plan for the Global Framework for Climate Services* (WMO, Geneva, 2014).
3 Vaughan, C. & Dessai, S. Climate services for society: origins, institutional arrangements, and design elements for an evaluation framework. *Wiley Interdisciplinary Reviews. Climate Change* **5**, 587–603, doi:10.1002/wcc.290 (2014).
4 Bircher, J. & Kuruvilla, S. Defining health by addressing individual, social, and environmental determinants: new opportunities for health care and public health. *Journal of Public Health Policy* **35**, 363–386, doi:10.1057/jphp.2014.19 (2014).
5 Barnes, C. S. *et al.* Climate change and our environment: the effect on respiratory and allergic disease. *The Journal of Allergy and Clinical Immunology in Practice* **1**, 137–141, doi:http://doi.org/10.1016/j.jaip.2012.07.002 (2013).
6 Mideksa, T. Economic and distributional impacts of climate change: the case of Ethiopia. *Global Environmental Change* **20**, 278–286 (2010).
7 Du, W., Fitzgerald, G. J., Clark, M. & Hou, X. Y. Health impacts of floods. *Prehospital and Disaster Medicine* **25**, 265–272 (2010).
8 Stanke, C., Kerac, M., Prudhomme, C., Medlock, J. & Murray, V. Health effects of drought: a systematic review of the evidence. *PLOS Currents Disasters*, doi:10.1371/currents.dis.7a2cee9e980f91ad7697b570bcc4b004 (2013).
9 Courtney, C. *Chris Courtney, The Nature of Disaster in China: The 1931 Yangzi River Flood* (Cambridge University Press, Cambridge, 2018).
10 Han, W., Liang, C., Jiang, B., Ma, W. & Zhang, Y. Major natural disasters in China, 1985–2014: occurrence and damages. *International Research and Public Healtth* **13**, 1118 (2016).
11 WB. *Natural Hazards Unnatural Disasters: Effective Prevention through and Economic Lens.* 231pp (World Bank, Washington, DC, 2010).

12 Coughlan de Perez, E. *et al.* Forecast-based financing: an approach for catalyzing humanitarian action based on extreme weather and climate forecasts. *Natural Hazards and Earth System Sciences* **15**, 895–904, doi:10.5194/nhess-15-895-2015 (2015).

13 CRED/UNISDR. *Disasters in Numbers.* 2pp (CRED/UNISDR, Brussels, 2015).

14 McGregor, G. R., Bessemoulin, P., Ebi, K. & Menne, B. *Heatwaves and Health: Guidance on Warning-System Development.* 114pp (WHO, Geneva, Switzerland, 2014).

15 Mason, S. J., Allen, T. & Thomson, M. C. in *Climate Information for Public Health Action* Ch. 4 (Routledge, London, 2018).

16 Gasparrini, A. *et al.* Mortality risk attributable to high and low ambient temperature: a multicountry observational study. *Lancet* **386**, 369–375, doi:dx.doi.org/10.1016/ (2015).

17 Wilhite, D. A. & Glantz, M. H. Understanding the drought phenomenon: the role of definitions. *Water International* **10** (3) 111–120 (1985).

18 Breban, R., Vardavas, R. & Blower, S. Theory versus data: how to calculate R0? *PloS One* **2**, e282, doi:doi.org/10.1371/journal.pone.0000282 (2007).

19 WHO-AFRO. *Mapping the Risk and Distribution of Epidemics in the WHO African Region: A Technical Report. 62* (WHO, AFRO, Geneva, Switzerland, 2017).

20 Molesworth, A. M., Cuevas, L. E., Connor, S. J., Morse, A. P. & Thomson, M. C. Environmental risk and meningitis epidemics in Africa. *Emerging Infectious Disease* **9**, 1287–1293 (2003).

21 McDermott, J. J., Staal, S. J., Freeman, A. H., Herrero, M. & Steeg, J. A. V. D. Sustaining intensification of smallholder livestock systems in the tropics. *Livestock Sciences* **130**, 95–109 (2010).

22 Krätli, S., Hülsebusch, C., Brooks, S. & Kaufmann, B. Pastoralism: a critical asset for food security under global climate change. *Animal Frontiers* **2**, 42–50 (2013).

23 Taylor, L. H., Latham, S. M. & Woolhouse, M. E. Risk factors for human disease emergence. *Philosophical Transactions of The Royal Society B Biological Sciences* **356**, 983–989 (2001).

24 Grace, D. *et al.* Mapping of poverty and likely zoonoses hotspots. Zoonoses Project 4. *Report to the UK Department for International Development. Nairobi, Kenya: ILRI* (2012).

25 Bett, B. *et al.* Effects of climate change on the occurrence and distribution of livestock diseases. *Preventive Veterinary Medicine* **137**, 119–129, doi:10.1016/j.prevetmed.2016.11.019 (2017).

26 Grace, D., Bett, B., Lindahl, J., & Robinson, T. *Climate and Livestock Disease: Assessing the Vulnerability of Agricultural Systems to Livestock Pests Under Climate Change Scenarios* (CCAFS, Copenhagen, Denmark, 2015).

27 Patz, J. A. & Hahn, M. B. Climate change and human health: a One Health approach. *Current Topics in Microbiology and Immunology* **366**, 141–171, doi:10.1007/82_2012_274 (2013).

28 WHO. *2017 Annual Review of Diseases Prioritized under the Research and Development Blueprint: Informal Consultation 24–25 January 2017* (WHO, Geneva, Switzerland, 2017).

29 Tirado, M. C. *et al.* Climate change and nutrition: creating a climate for nutrition security. *Food and Nutrition Bulletin* **34**, 533–547 (2013).

30 UNICEF. *Improving Child Nutrition: The Achievable Imperative for Global Progress* (UNICEF, New York, 2013).

31 Regan, L., Bailey, R. L. & Black, R. E. The epidemiology of global micronutrient deficiencies. *Annals of Nutrition and Metabolism* **66**, 22–33 (2015).

32 Suskind, R. M. & Varma, R. N. Assessment of nutritional status of children. *Pediatrics in Review* **5**, 195–202, doi:10.1542/pir.5-7-195 (1984).

33 Ramakrishnan, U. Nutrition and low birth weight: from research to practice. *American Journal of Clinical Nutrition* **79**, 17–21 (2004).

34 Brabin, B. J., Agbaje, S. O. F., Ahmed, Y. & Briggs, N. D. A birthweight nomogram for Africa, as a malaria-control indicator. *Annals of the Tropical Medicine and Parasitology* **93**, S43–S57 (1999).

35 Wort, U. U., Hastings, I. M., Carlstedt, A., Mutabingwa, T. K. & Brabin, B. J. Impact of El Niño and malaria on birthweight in two areas of Tanzania with different malaria transmission patterns. *International Journal of Epidemilogy* **33**, 1311–1319 (2004).

36 Magan, N., Medina, A. & Aldred, D. REVIEW: Possible climate-change effects on mycotoxin contamination of food crops pre- and postharvest. *Plant Pathology* **60**, 150–163, doi:10.1111/j.1365-3059.2010.02412.x (2011).
37 Myers, S. S. *et al.* Increasing CO2 threatens human nutrition. *Nature* **510**, 139–142, doi:10.1038/nature13179 (2014).
38 Headey, D. & Shenggen, F. Anatomy of a crisis: the causes and consequences of surging food prices. *Agricultural Economics* **39**, 375–391 (2008).
39 Torlesse, H. L., Kiess, M. & Bloem, W. Association of household rice expenditure with child nutritional status indicates a role for macroeconomic food policy in combating malnutrition. *Community and International Nutrition* **133**, 1320 (2003).
40 FAOSTAT. FAOSTAT data – Food Balance (2017).
41 FAOSTAT. FAOSTAT data – Production (2017).
42 DeFries, R. *et al.* Synergies and trade-offs for sustainable agriculture: nutritional yields and climate-resilience for cereal crops in central India. *Global Food Security* **11**, 44–53 (2016).
43 DeFries, R. *et al.* (2015) Metrics for land-scarce agriculture: nutrient content must be better integrated into planning. *Science* **349**, 238–240 (2015).
44 Mupunga, I., Mngqawa, P. & Katerere, D. R. Peanuts, aflatoxins and undernutrition in children in Sub-Saharan Africa. *Nutrients* **9**, 1287, doi:10.3390/nu9121287 (2017).
45 Smith, L. E. *et al.* Examining environmental drivers of spatial variability in Aflotoxin accumulation in Kenyan Maize: potential utility in risk prediction models. *African Journal of Food, Agriculture, Nutrition, and Development* **16**, 11086–11105, doi:10.18697/ajfand.75.ILRI09 (2016).
46 Tirado, M. C., Clarke, R., Jaykus, L. A., McQuatters-Gollop, A. & Franke, M. J. Climate change and food safety: a review. *Food Research International* **43**, 1745–1765, doi:10.1016/j.foodres.2010.07.003 (2010).
47 Dushoff, J., Plotkin, J. B., Levin, S. A. & Earn, D. J. Dynamical resonance can account for seasonality of influenza epidemics. *Proceedings of the National Academies of Science* **101**, 16915–16916 (2004).
48 Devereux, S., Sabates-Wheeler, R. & Longhurst, R. *Seasonality, Rural Livelihoods and Development*, 334 (Routledge, London, 2011).
49 Kelly-Hope, L. A. & Thomson, M. C. in *Seasonal Forecasts, Climatic Change, and Human Health* (eds. M.C. Thomson, R. Garcia-Herrera, & M. Beniston), 31–70 (Springer Science+Business Media, Berlin, Germany, 2008).
50 Shaman, J., Pitzer, V. E., Viboud, C., Grenfell, B. T. & Lipsitch, M. Absolute humidity and the seasonal onset of influenza in the continental United States. *PLOS Biology* **8**, e1000316, doi:10.1371/journal.pbio.1000316 (2010).
51 Lolekha, S. *et al.* Effect of climatic factors and population density on varicella zoster virus epidemiology within a tropical country *American Journal of Tropical Medicine and Hygiene* **64**, 131–136 (2001).
52 Perez Garcia-Pando, C. *et al.* Soil dust aerosols and wind as predictors of seasonal meningitis incidence in Niger. *Environmental and Health Perspectives* **122**, 679–686 (2014).
53 Karkey, A. *et al.* The ecological dynamics of fecal contamination and Salmonella Typhi and Salmonella Paratyphi A in municipal Kathmandu drinking water. *PLOS Neglected Tropical Diseases* **10**, e0004346 (2016).
54 Koelle, K., Rodo, X., Pascual, M., Yunus, M. & Mostafa, G. Refractory periods and climate forcing in cholera dynamics. *Nature* **436**, 696–700 (2005).
55 Rogers, D. & Randolph, S. Climate change and vector-borne diseases. *Advances in Parasitology* **62**, 345–381 (2006).
56 Luis, A. D., Douglass, R. J., Mills, J. N. & Bjørnstad, O. N. The effect of seasonality, density and climate on the population dynamics of Montana deer mice, important reservoir hosts for Sin Nombre hantavirus. *Journal of Animal Ecology* **79**, 462–470 (2010).
57 Weinberger, D. M., Klugman, K. P., Steiner, C. A., Simonsen, L. & Viboud, C. Association between Respiratory Syncytial Virus activity and pneumococcal disease in infants: A time series analysis of US hospitalization data. *PLOS Mediciine* **12**, e1001776 (2015).

58 Stevenson, T. J. *et al.* Disrupted seasonal biology impacts health, food security and ecosystems. *Proceedings of the Royal Society B: Biological Sciences* **282**, 20151453, doi:10.1098/rspb.2015.1453 (2015).
59 Paynter, S. *et al.* Using mathematical transmission modelling to investigate drivers of respiratory syncytial virus seasonality in children in the Philippines. *PloS One* **9**, e90094 (2014).
60 Martinez-Bakker, M., Bakker, K. M., King, A. A. & Rohani, P. Human birth seasonality: latitudinal gradient and interplay with childhood disease dynamics. *Proceedings of the Royal Society of London: Biological Sciences* **281**, 20132438 (2014).
61 Currie, J. & Schwandt, H. Within-mother analysis of seasonal patterns in health at birth. *Proceedings of the Nattional Academy of Sciences* **110**, 12265–12270 (2013).
62 Bjørnstad, O., Finkenstadt, N. B. & Grenfell, B. T. Endemic and epidemic dynamics of measles: Estimating epidemiological scaling with a time series SIR model. *Ecological Monographs* **72**, 169–184 (2002).
63 Rain, D. *Eaters of the Dry Season: Circular Labor Migration in the West African Sahel* (Westview Press, Boulder, CO, 1999).
64 Ferrari, M. J. *et al.* The dynamics of measles in sub-Saharan Africa. *Nature* **451**, 679–684 (2008).
65 Bharti, N. *et al.* Spatial dynamics of meningococcal meningitis in Niger: observed patterns in comparison with measles. *Epidemiology and infection* **140**, 1356–1365 (2012).
66 Wesolowski, A. *et al.* Multinational patterns of seasonal asymmetry in human movement influence infectious disease dynamics. *Nature Communications* **8,** 2069 (2017).
67 Metcalf, C. J. E. *et al.* The seasonal and climatic determinants of access to care: implications for measles outbreak risk in Madagascar. *Lancet* **389**, S14 (2017).
68 Moore, S. E. *et al.* Season of birth predicts mortality in rural Gambia. *Nature* **388,** 434 (1997).
69 James, P., Silver, M. & Prentice, A. Epigenetics, nutrition, and infant health in *The Biology of the First 1,000 Days* (eds. Karakochuk, C. D., Whitfield, K. C., Green, T. J., & Kraemer, K.), 335–353 (CRC Press, Boca Raton, FL, 2018).
70 Batterbury, S. & Warren, A. The African Sahel 25 years after the great drought: assessing progress and moving towards new agendas and approaches. *Global Environmental Change* **11**, 1–8 (2001).
71 Giannini, A., R. Saravanan, R. & Chang, P. Oceanic forcing of Sahel rainfall on interannual to interdecadal time scales. *Science* **302**, 1027–1030 (2003).
72 Booth, B. B. B., Dunstone, N. J., Halloran, P. R., Andrews, T. & Bellouin, N. Aerosols implicated as a prime driver of twentieth-century North Atlantic climate variability. *Nature Letter* **484**, 228–232, doi:10.1038/nature10946 (2012).
73 Mouchet, J., Faye, O., Julvez, J. & Manguin, S. Drought and malaria retreat in the Sahel, West Africa. *Lancet* **348**, 1735–1736 (1996).
74 McCann, J. C. Climate and causation in African History. *The International Journal of African Historical Studies* **32**, 261–279 (1999).
75 Nour, N. N. Maternal health considerations during disaster relief. *Review of Obstetrics and Gynecology* **4**, 22–27 (2011).
76 Moreno-Walton, L. & Koenig, K. Disaster resilience: addressing gender disparities. *World Medical and Health Policy* **8**, 46–57 (2016).
77 Gamble, J. L. *et al.* Climate change and older Americans: state of the science. *Environmental Health Perspectives* **121**, 15–22, doi:10.1289/ehp.1205223 (2013).
78 Dong, Q. Seasonal changes and seasonal regimen in Hippocrates. *Journal of Cambridge Studies* **6,** 128 (2011).

3

CONNECTING CLIMATE INFORMATION WITH HEALTH OUTCOMES

Madeleine C. Thomson, C. Jessica E. Metcalf and Simon J. Mason
Contributors: Adrian M. Tompkins and Mary Hayden

Happy the man who has been able to learn the causes of things

Georgics, II *by Virgil c. 29BC*

3.1 Introduction

In agricultural development water is understood as a precious and finite resource that must be used wisely to maximize crop growth. The relationship between available water and crop growth is dependent on soil type, plant cultivar and development stage, and is relatively easy to calculate; after all, the plant stays in the same place throughout its growing period. In contrast, pathogens and animal and human hosts exhibit a complex set of biological and behavioural interactions in response to climatic and environmental drivers that may vary in both time and space. As climate may be a significant driver of a wide range of health outcomes (see Chapter 2), climate information can potentially support a wide range of health decisions (see Box 3.1).

All decision processes involved in the prevention or control of climate-sensitive health issues (such as the decision to spray houses with indoor residual insecticides for the control of malaria-bearing mosquitoes) have their own spatial or temporal context. Understanding this context, including climatic, environmental and population characteristics, is the first step to using climate information effectively. The spatial or temporal dynamics of the problem as well as potential solution(s) are key issues in decision-making processes. For instance, heat waves may be of particular concern to elderly populations in urban environments. Prevention/control decisions for heat waves may be made at the individual level, the local administrative level or may require a regional or national process. They may be routine control efforts enacted prior to the seasonal occurrence of heat waves. Alternatively, control

BOX 3.1. CLIMATE INFORMATION OPPORTUNITIES

Climate information has the potential to inform a wide range of health decisions[1] through an improved understanding of the following:

- **Mechanisms of Disease Transmission:** to help identify new opportunities for intervention.
- **Spatial Risk:** to help identify populations at risk for better targeting of interventions.
- **Seasonal Risk:** to inform the timing of routine interventions.
- **Sub-seasonal and Year-to-Year Changes in Risk:** to identify when changes in epidemic risk are likely to occur to initiate appropriate prevention and response strategies.
- **Trends in Risk:** to identify long-term drivers of disease occurrence (including shifts in the climate) to plan for and support future prevention and response strategies.
- **Assessment of the Impacts of Interventions:** to remove the role of climate if it interferes with the proper assessment of interventions.

efforts may be initiated in response to occasional extreme events that visit a region during an unusual year. Changes in control strategies (including urban planning) may also be undertaken in response to observed shifts in the underlying frequency and intensity of heat waves.

In this chapter we consider a range of factors that need to be taken into account when seeking to use climate information to improve health decision-making. Identifying causal mechanisms that link climate drivers with specific health issues is an important starting point for policy-makers. Matching decision time-horizons to climate information in a way that takes account of scale issues, uncertainties in the underlying data and modelling approaches as well as institutional barriers to knowledge and data sharing is also critical. And of course, all of this is dependent on a solid understanding of the climate information (including its limitations) that is available to health decision-makers. A researcher may be satisfied with a simple times-series of climate data from an authoritative source; a decision-maker needs to know that the climate information is robust, available for routine use and scalable (i.e., can be used over the entire region of interest).

3.2 Climate information for use in health decision-making

Climate is measured routinely around the world by the meteorological/climate community using internationally agreed standards, with a significant amount of data shared freely in real-time (§ 6.2). Historical and current climate data and information products (§§ 6.3 and 6.4) can be used to inform a wide range of planning processes as well as in early warning systems (EWS). Information about the future weather and/or climate varies in specificity (lead-time, spatial and temporal averaging), the regions and seasons where it is most accurate and its status in terms of operational delivery. Information that is currently available across the globe for routine operational decision-making can be divided into three specific timeframes: weather, season and climate change. These timescales are related to spatial scale: at the one extreme, short-term weather predictions are reliable at the local level, while, at the other, climate change trends (especially temperature) are most reliable at the subcontinental level. Climate change scenarios, indicating possible long-term changes in temperature and rainfall, provide important guidance to climate change *mitigation*, motivating reductions in carbon emissions from the health sector itself and promoting the health co-benefits of a low-carbon economy. When it comes to *adapting* to climate variability and change, information about possible future climates is important for planning major infrastructure developments or considering long-term policy shifts (e.g., from malaria control to malaria eradication[2]). However, most health programming decisions are made at seasonal to annual timescales or respond to four- to five-year political or funding cycles.[3] Timescales of climate information need to be matched with these time horizons of decision-making (Table 3.1).

Weather can be predicted at the local scale for several days ahead with a reasonable level of accuracy (particularly in the extratropics), but the accuracy after three to five days has deteriorated considerably (see Chapter 7). The seasonal climate

TABLE 3.1 Time horizons for decision-making in the health sector

Investment 2–5 decades	Carbon emissions mitigation strategies Malaria eradication strategies Major infrastructure investment Workforce development
Strategic planning 6–20 years	Research and development of medical countermeasures (e.g., drugs, vaccines) and vector control tools (e.g., new insecticides) Improved nutritional content of crops Health facility investments Curriculum development
Policy cycles 2–5 years	4- to 5-year political cycle Health service re-organization 2- to 5-year research grant cycle
Planning cycles < 2yrs	Annual planning and commissioning cycle Demand for visible 'quick wins' from funders
Seasonal preparedness and response < 4 months	Seasonal planning cycle Epidemic/disaster preparedness and response
Weekly facility management < 1 week	Weather disaster preparedness and response Patient scheduling for non-urgent cases

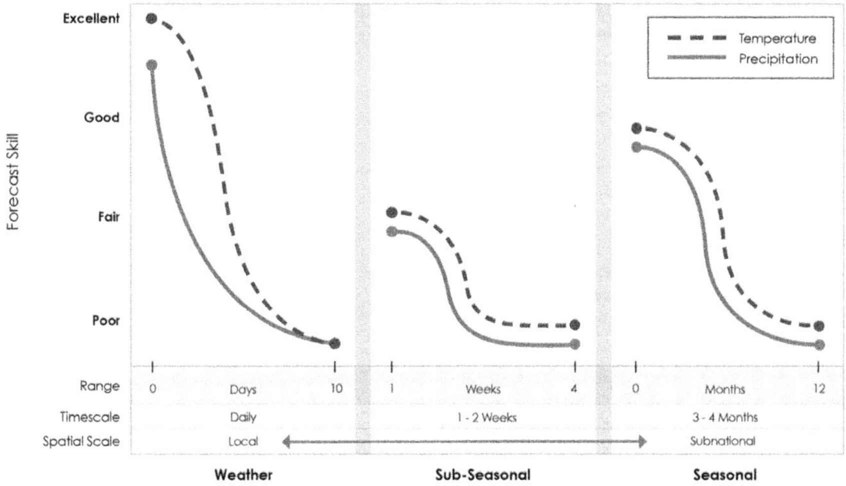

FIGURE 3.1 Best forecast skill at multiple timescales with indications of the forecast ranges, timescales and spatial scales over which the forecasts are averaged

may be affected by phenomena such as the El Niño – Southern Oscillation (ENSO; see Box 5.1), giving advanced warning of unusual rainfall or temperature months in advance.The ENSO and other influences underpin seasonal climate forecasts, which are most robust in the tropics (see Chapter 8). Projections of trends in 30-year averaged climate due to greenhouse gas emissions are available for the long-term and are considered robust for temperature (see Chapter 9). Three other timeframes are only briefly considered in this book because they remain, for now, substantially in the research arena, namely sub-seasonal, multi-annual and decadal. Emerging capacities in forecast capability at sub-seasonal timescales (e.g., seven-day averaged weather two weeks or more in advance) and multi-annual timescales provide new opportunities for health research, but evidence to date of potential utility is for specific locations only. Long sought for decadal prediction (five- to ten-year averaged climate going out over ten- to 30-year timescales) is a focus of intense research. A schematic representation of predictability of anomalies at the shorter verifiable timescales is presented in Figure 3.1.

3.3 Data issues

In order to facilitate climate and health analyses, data must be shared between communities. Data sharing is an issue in nearly every organization because there are consequences, both good and bad, to sharing information beyond institutional borders. There are many barriers to sharing health data, even within the health community (see Table 3.2),[4] and these barriers can be even greater when it comes to sharing data between different sectors, such as climate and health. There are also significant barriers to the sharing of climate data, especially observations from meteorological stations at high temporal resolution – e.g., daily data (see §§ 6.4.1 and Box 6.1).

TABLE 3.2 Evidence for barriers to sharing of routinely collected public health data[4]

Technical
1. Data not collected
2. Data not preserved
3. Data not found
4. Language barrier
5. Restrictive data format
6. Technical solutions not available
7. Lack of metadata and standards
Motivational
8. No incentives
9. Opportunity cost
10. Possible criticism
11. Disagreement on data use
Economic
12. Possible economic damage
13. Lack of resources
Political
14. Lack of trust
15. Restrictive policies
16. Lack of guidelines
Legal
17. Ownership and copyright
18. Protection of privacy
Ethical
19. Lack of proportionality
20. Lack of reciprocity

Increasingly, countries are developing Open Data policies, where government information is made visible and available. The main goal is to harness the data revolution for sustainable development[5] with a focus on climate, health and agriculture. Open data policies will take time to transform data culture and improve data sharing capabilities at the national level. Since the direction towards greater openness is already underway, improving the capacity of health practitioners and researchers to use these new data sources effectively is a critical step that needs to be addressed.

How data are interpreted will vary according to the knowledge and experience of the individual user and the way the information is presented. Maps of likely hotspots or regions at risk provide a simple visual tool for decision-makers. However, all maps simplify reality and, because of this, learning to read such maps and understand the information that has been emphasized or neglected in their creation is important in order to make valid inferences. Trust is at the heart of information uptake. Ensuring that sources of data are authoritative and provided with associated meta-data (a set of data that describes and gives information about the data being considered) is key. A healthy scepticism is a valuable asset when exploring new and unfamiliar data sources.

3.4 Exploring relationships

The impact of climate and weather on health is often not immediate. Even when deaths occur from drownings associated with unusually heavy rains there will likely be a delay of hours to days between rain falling and floods occurring – as water takes time to move down rivers and tributaries.

For a vector-borne disease such as Zika, the population dynamics of the vectors (*Aedes aegypti* and *Ae. Albopictus)* and virus need to be taken into account when exploring lags between climate drivers and health indicators. In addition to delays associated with vector and pathogen dynamics, the development of a seasonal or epidemic wave is largely attributable to the changing proportion of susceptible hosts in the population (see Figure 3.2). Further, lags in the relationship of climate and health outcomes may be attributed to delays in the manifestation of the disease – e.g., if the disease impacts on the foetus in utero and the child is only included as a case after birth. Manifestations of symptoms associated with Zika virus infection in Brazil, including acute rash, Guillain-Barré syndrome and suspected microcephaly, peaked during epidemiological week 17, 26 and 48 respectively.[6] It is these transmission lags that allow the creation of EWS based on current and historical environmental and climatic data. For vector-borne diseases in locations subject to distinct rainy seasons, the lag between peak rainfall season and peak cases of disease is commonly around two to three months although the duration of the lag will depend on the climatic conditions[7] including the distribution and intensity of rainfall, as well as temperature and humidity.

The relationship between temperature and disease transmission is even more complex.[8] The impact of temperature on the development rates of organisms is amenable to laboratory as well as observational studies. The basic biological response follows a thermal response curve, i.e., has a lower bound minimum, an optimal temperature and a higher bound maximum. This curve may be estimated for a number of different physiological processes occurring in the pathogen, the

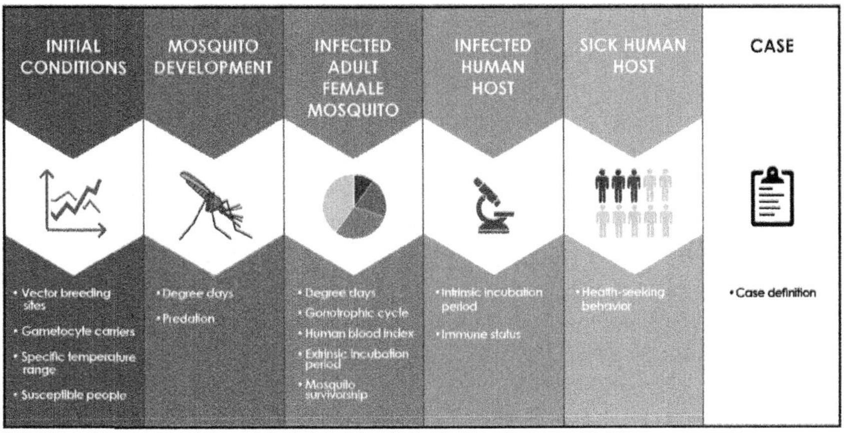

FIGURE 3.2 Understanding lags between climatic events and cases of disease

vector or the human host and can be compared with field observations.[9] Not all vector-borne diseases favour warmer climates. Transmission of bubonic plague occurs in cooler mountainous regions (see Case Study 3.1).

CASE STUDY 3.1 PLAGUE, RETURN OF AN OLD FOE

Mary Hayden, National Center for Atmospheric Research, Boulder, USA

Plague (*Yersinia pestis*) is a bacterial disease that has caused pandemics that have literally changed the course of history; the infamous Black Death, which occurred in the mid-1300s and wiped out a third of Europe's population is one example. Rats and the fleas that they carry have long been viewed as the main sources of human infection resulting in *bubonic* plague. Human cases have re-emerged in recent years as a result of changing environments and weak or non-existent surveillance.[10] Human disease usually occurs in one of two forms; *bubonic* (typically dependent on transmission by fleas) and *pneumonic* plague which often occurs when bubonic plague victims are not treated and the infection travels to the lungs. Once in the lungs, the disease is spread from person to person through respiratory droplets. More than 90% of today's plague cases occur in Africa.[i] Plague is commonly found in cooler highland environments where lower temperatures increase the likelihood of transmission to rat or human hosts by the most common vector, the flea *Xenopsylla cheopis*. Transmission by fleas to rats or humans occurs when the ingested blood meal in their stomach coagulates at temperatures below 27°C following the activation of a coagulase enzyme. *Y. pestis* bacteria, which are ingested with the bloodmeal, are able to multiply in the blood clots which are then regurgitated when the flea next takes a bite – allowing the bacteria to penetrate the bite wound and infect the bitten rat or person. At temperatures above 27°C coagulase is not produced, and the blood meal does not coagulate; *Y. perstis* passes through the flea gut and is not regurgitated into the bite wound. As a result, bubonic plague epidemics are not common in environments where temperatures reach above 27°C.

The West Nile region in northwestern Uganda is a focal point for human plague, which peaks in boreal autumn after the main rainy season.[11, 12] The United States Centers for Disease Control and Prevention (CDC) partnered with the National Center for Atmospheric Research (NCAR) to address the linkages between climate and human plague risk in this region in order to develop a better understanding of potential control options. Because *in-situ* meteorological records are sparse, a hybrid dynamical–statistical meteorological downscaling technique was applied to generate a multi-year high spatial resolution climate dataset based on NCAR's Weather Research and Forecasting Model.[13] The dataset was subsequently employed to develop a spatial risk model for human plague occurrence in the West Nile region above 1300 meters, which is cooler and wetter than surrounding areas.[14]

Further complicating the statistical analysis of these relationships, temperature variations are often correlated with those of rainfall (see § 5.3.5).

Given the complex interactions between temperature and malaria transmission it is hard to specify which temperature sensitive aspect of transmission is most important in establishing this lag. As mentioned above it is these lags between observed climatic variables (see § 4.2) and disease indicators that provide the opportunity for the development of climate-informed EWS. When rainfall or temperature changes are predictable, then additional time can be added to the EWS with the use of weather and climate forecasts (see Chapters 7 and 8).

Although for some health outcomes and contexts there may be a strong, linear relationship between a climate driver and cases, the relationship is often highly non-linear. For example, low levels of rainfall (< 2.5 mm/day) in Botswana appear to have a near-linear relationship to anomalies in malaria cases but a quadratic relationship is clearly observed when wetter conditions are also included in the analysis (Figure 3.3).

The decline in malaria at higher rainfall levels is often attributed to the washing out of mosquito breeding sites during heavy rains. A statistical model developed using moderate to low rainfall years alone would have performed poorly in out-of-sample very wet years. In the Botswana example, the most extreme rainfall was associated with cyclone Eline in 2000, a data point that was not used to generate the model.[15] While heavy rainfall may destroy vector breeding it may create new sites at the end of the rainy season when flood waters retreat. A modelling approach which takes into account the variations in the seasonality of transmission and disease incidence will better capture on-going processes.[16]

An important source of non-linearity in climate–health interactions is the immune response to infection which may protect survivors from re-infection or

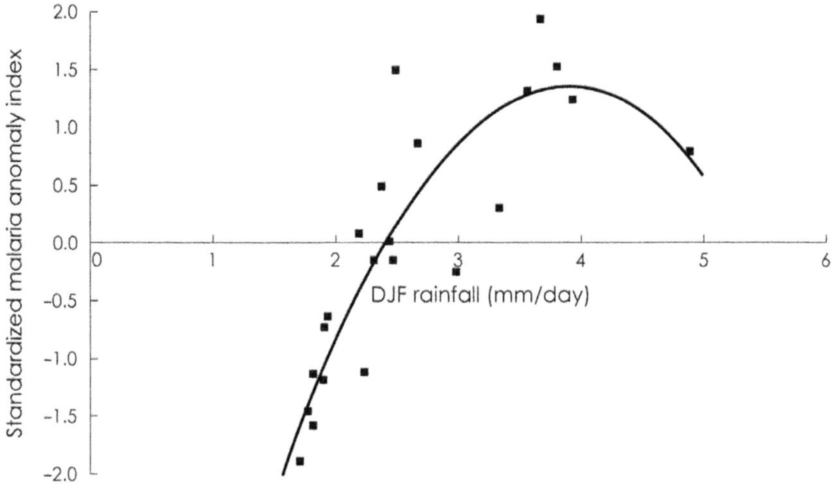

FIGURE 3.3 Relationship between annual malaria anomalies and December to January rainfall in Botswana

disease for a period of time. The immune status of a population can quickly cause significant changes in the proportion of the population susceptible to infection, and explains why epidemics of infectious diseases often peak and rapidly decline once the source of susceptible individuals is significantly reduced. As higher rates of immunity are commonly found in regions with long or year-round transmission seasons, the impact of climate variability on cases may be effectively buffered.[17] Statistical models that are based on the assumption that the underlying susceptible population is constant will under-estimate the climate's relationship to disease. Non-linearities in the relationship of climate drivers to disease outcomes are an important reason for considering the use of mathematical models in climate disease analysis as they are able to capture some of these dynamics.

3.5 Linking climate to health outcomes

A model attempts to link climate, and possibly other drivers (the model input), to the targeted health outcome (the model output). If the mechanisms and processes that link the two are poorly understood, statistical fitting or 'machine learning' techniques can still be used to provide this link. Alternatively, if a good understanding of the biological processes that drive the health outcome is available, numerical (mechanistic) models can be derived (§ 8.2.2). As numerical models often make use of statistical approximations when aspects of the model are unknown (just like the parameterizations in weather and climate models; § 7.4.4), it might be best to consider that these two approaches lie upon a continuum.[18] The applicability of the approach chosen will depend on characteristics of the pathogen and the host–pathogen relationship as well as the availability of data and information on underlying mechanisms. Of course, it is impossible for the complex systems that we are dealing with (climate and disease) to be *exactly* represented by any simple model. However, the approximations made by well-chosen models can be extremely useful.

3.5.1 Statistical models

Statistical models linking health outcomes and climate exposures, such as climate extremes, varying lengths or intensities of the rainy season, or trends in minimum temperature, can give indications of underlying mechanisms worthy of further exploration, but cannot be used to definitively establish causation because of the problem of confounding variables. Climate may impact, or simply be correlated with, other processes or variables that the researcher is unaware of, and as a consequence the researcher may infer a relationship when there is none. Randomized Controlled Trials (RCTs) are the gold standard study design for the evaluation of medical interventions because they can effectively control for confounding variables. In RCTs participants are randomly allocated between groups to minimize systematic differences between control and interventions groups, and associated biases that might result. In contrast, public health research, including that associated with climate, relies primarily on observational methods.[19] Because the research is often conducted at the population level, it is usually not possible to fully randomize the exposed population on cost, practicality or ethical grounds. As an example, road accidents are positively correlated with monthly temperatures in Europe simply because more and longer journeys are made in the summer months, even though inclement winter weather obviously can cause accidents. In this case, including a denominator in the analysis (number of miles travelled and/or number of drivers on the roads) would reveal the weather-related accident risk to individual drivers.

Systematic reviews from specialist organizations are increasingly used to provide objective and transparent evidence from both RCTs and observational studies (Box 3.2). A systematic approach to reviewing evidence can also inform climate risk, adaptation and mitigation strategies. However unique challenges exist in terms of integrating disparate data as well as analytical norms of different communities. This complexity is illustrated by a systematic review of drought impacts on health.[20] The review concluded: '*The probability of drought-related health impacts varies widely and largely depends upon drought severity, baseline population vulnerability, existing health and sanitation infrastructure, and available resources with which to mitigate impacts as they occur.*'

Climate and health is an emerging field and there may be insufficient peer reviewed literature for a systematic review. Under these circumstances a Delphi review,[21] where the collective opinion of a group of experts is accessed using a structured process, may be used instead. This approach to critical review is based on the premise that intelligence from a pool of experts can enhance individual judgement if expressed independently. The potential biases associated with observational studies mean that researchers undertaking such studies must pay particular attention to the plausibility of relationships observed. Before using a statistical model to

BOX 3.2 USE OF SYSTEMATIC REVIEWS

The International Cochrane Collaboration (ICC)[ii] produces systematic reviews of primary research (usually RCTs) in human health care and health policy and is recognized as the highest standard in evidence-based health care. The ICC investigates the effects of health interventions for prevention, treatment and rehabilitation as well as the effects of diagnostic tests under specific conditions. Other organizations that produce rigorous systematic reviews are: the Campbell Collaboration,[iii] that produces and disseminates systematic reviews on the effects of interventions in the social and behavioural sciences; the National Institute for Health and Clinical Excellence (NICE),[iv] which commissions systematic reviews on new and existing technologies and then uses them to make recommendations to the UK's National Health Service, and the Centre for Reviews and Dissemination (CRD),[v] which produces systematic reviews of health interventions.

create a forecast, a mechanism by which the observed association might be considered plausible is needed. A number of critical questions that should be asked when developing a decision-support model are proposed in Box 3.3.

BOX 3.3 FIVE QUESTIONS TO CONSIDER WHEN DEVELOPING CLIMATE-DRIVEN MODELS FOR DECISION-SUPPORT (ADAPTED FROM[22])

Q1: Do you understand the underlying data?
Make sure the data used, as well as the processes that generated it, are of the highest possible quality, and that you fully understand them.

Q2: Will the model results will be meaningful to a decision-maker?
Good models usually tell a clear story. If the models you're using don't give you one they may need to be refined. Complex models may be inevitable but they still need to be thought through, refined and simplified enough to make them understandable to those that need to use them.

Q3: Is the model as simple as possible – but not simpler (Occam's Razor)?
Predictability typically first improves and then deteriorates as model complexity increases, so adding complexity should not be a goal in itself. Of course, there are also risks to oversimplifying the model, so a judgement needs to be made. Einstein is often quoted as saying, '*Everything should be as simple as it can be, but not simpler*'; a good principle to apply to predictive analytics.

Q4: Have you tested the predictive accuracy of the model using independent data and across multiple similar environments?
If the results look to good to be true it's often an over-fitting issue; a common error in predictive analytics. Over-fitting means that the model is too strongly tailored to the data that was used in its development (the training data). Over-fitted models do not predict well new data and therefore cannot make good forecast models. To avoid over-fitting apply the model to fresh data (e.g., through cross validation) in new, but similar, contexts while the model is being developed. A good predictive model should be nearly as accurate with new data as it is with the training set.

Q5: Is the model still relevant?
Models that have worked well in the past may no longer be relevant (think about economic forecasts!). Because it takes time and energy to develop a model it may be easy to see predictability where there is none. If the data don't support your predictions, you should be prepared to jettison your model – possibly multiple times. Good models are developed through deep understanding of the context for the model development and a very honest interpretation of results.

3.5.2 Mathematical models

Mathematical models of an infectious disease are generally framed around flows between core categories of individuals (e.g., susceptible, exposed, infected, recovered) and sometimes vectors via equations that assume causation and may encompass process uncertainty, climate drivers, etc. An array of options is available to link these mechanistic frameworks to time-series of disease incidence data for a focal infectious disease, e.g., basic maximum likelihood approaches,[23] Markov Chain Monte Carlo or iterated filtering-type approaches.[24] These framings can then be used to test hypotheses about important drivers via comparison of how well the model outputs fit the predictand[25] and can be used to forecast future incidence modulated by a changing climate. Thus, numerical modelling can contribute to understanding the past, and in particular disentangling the relative role of core drivers as well as predicting the future. For infectious disease models, the goal has frequently been to explore different interventions scenarios in order to inform priority-setting for policy-makers.[26] However, in recent years there is increasing interest in using models for real-time forecasting,[27] although there remains a significant gap in the operational readiness of the numerous forecasting systems presented in the literature.[28]

CASE STUDY 3.2 SOURCES OF UNCERTAINTY IN MODELLING CLIMATE AND MALARIA

Adrian M. Tompkins, Abdus Salam International Centre for Theoretical Physics, Trieste, Italy

The relative importance of transmission model uncertainty, initial condition uncertainty and driving climate uncertainty (Figure 3.4) has been explored using meteorological and malaria data from a highland tea plantation in Kericho, Kenya situated close to the temperature threshold for transmission.[29] A genetic algorithm was used to calibrate each of these three factors within their assessed prior uncertainty in turn to see which allowed the best fit to a time series of approximately 25 years of confirmed malaria cases (the predictand). The spatial representativeness uncertainty for temperature dominated the uncertainty due to model parameter settings. Initial condition uncertainty played a little role after the first two years and is thus important in the EWS context, but negligible for decadal and climate change investigations. Thus, while reducing uncertainty in the model parameters would improve the quality of the simulations, the uncertainty in the temperature data is critical (see Chapter 6). This result is a function of the mean climate of the location itself and model uncertainty would be relatively more important at warmer, lower altitude locations. Uncertainty in model development is then compounded by uncertainty in the way model outputs are used – either for furthering research or supporting decisions (Figure 3.4).

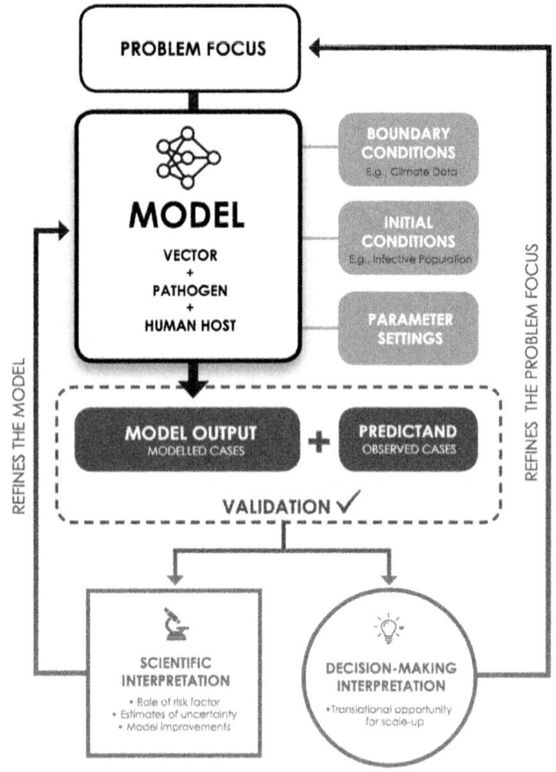

FIGURE 3.4 Schematic of the potential sources of uncertainty when using a weather/climate-sensitive disease model to simulate observed health outcomes

Because models can only be an approximation of the truth, statisticians and mathematicians have developed modelling tools that can express key elements of the uncertainty in the model. In deterministic models, the output of the model is decided by the parameter values and the initial conditions. However many times you run the model with the same inputs you will get the same result. A stochastic model is a tool used for accounting for known uncertainties in one or more of the parameter values by allowing for random variation over time. The same set of parameter values and initial conditions will lead to an ensemble of different outputs (see Box 7.6). Stochastic models possess some inherent randomness and are used in climate science as well as in disease modelling as they provide an assessment of the range of likely outcomes.

Policy-makers and practitioners need to know how certain the data are that they are using to drive decisions. Understanding sources of uncertainty in the development of climate-driven disease forecasts helps decision-makers understand where, when and why forecasts may be more or less robust and allows the prioritization of error reduction.

When using a weather-sensitive disease model to simulate observed health outcomes, uncertainty may derive from a variety of sources including the driving climate/environmental data (termed boundary conditions), the entomological and epidemiological initial state (termed initial conditions), the model structure and parameter settings, and lastly errors in the health data itself (the predictand) (see Case Study 3.2).

Aspects of uncertainty are inadequately considered in much epidemiological research[30] but are usually considered in climate modelling.[31] A common approach taken by the climate community in the development of seasonal climate forecasts is to employ an ensemble of models to create a probability distribution of possible outcomes (see § 7.4). The basic idea is that biases due to imperfect models will tend to cancel each other out. Sometimes probability distributions will be broad and there will be little predictability in the system. However, where there is a sharp probability distribution, predictability is stronger (assuming that the ensemble forecast system is well-calibrated) and the information may be used by decision-makers for taking precautionary action.[32] The main reason for using a probabilistic system is that users should not be misled by overconfident forecasts. Multi-model ensemble approaches are increasingly being used in the development of health EWS.[32–34] However, a common challenge to weather/climate and health forecasting is that information on the current situation is required for model initialization[35] and relevant epidemiological and entomological information is rarely available in near real-time, if at all.

3.6 Working with uncertain forecasts

Working with uncertain information is a decision-making problem: the forecaster's job is to try to quantify and minimize the uncertainty in the level of risk and timing of a hazard or event, while it is the practitioner's job to manage that risk and the uncertainty associated with it. Questions of how to manage uncertainties occur in all walks of life, including in public health management: when will the next flu epidemic occur, for example, and how many people will be affected? In Chapter 6, the availability of climate data to estimate public health risk is discussed in detail. Such information is useful for knowing which hazards to worry about at which time of the year, and perhaps how the risks have changed over the last few years. However, ideally we would like to know what the risks are in the coming days (Chapter 7), weeks or season (Chapter 8) or years to decades (Chapter 9), and how they differ from what might be considered normal.

There are some hazards that we have to be prepared for all the time, such as earthquakes, which have no seasonal pattern and can strike without warning. However, it could be exceptionally inefficient if we were having to worry constantly about whether there is likely to be a major storm tomorrow simply because it is the middle of the wet season right now. Forecasts reduce the uncertainty in the risk, making management of that risk easier (but not necessarily easy). Knowing the accuracy or reliability of the forecasts is a prerequisite to identifying the best ways of managing the risk (the distinction between accurate and reliable forecasts is explained in Box 7.4).

Probabilistic forecasts (see Box 7.5) can maximize the time available to prepare, while minimizing the risk of a false alarm to a level considered acceptable by decision-makers. In the United Kingdom, heat alerts are triggered when there is a 60% probability of critical day- and night-time temperature thresholds being reached on at least two consecutive days. UK forecasts usually reach the minimum 60% confidence level two to three days before a heat wave hits, but when a confident forecast is achieved with a longer lead-time, an alert could be given earlier.[36]

Assessing the effectiveness of a climate-informed intervention is more problematic than assessing the direct impact of the climate event alone. At its most basic, evidence of the utility of a EWS requires that morbidity and mortality from a predicted event are compared with a realistic assessment of the hypothetical outcome if the early warning intervention had not been in place. Put another way, one must be able to discern between EWS 'false alarms' and non-occurring 'epidemics' that were prevented by timely action based on the system. A comparison between what actually happened and what would have happened in the absence of the intervention is known as a counter-factual analysis. The simplest approach is to compare the impact of a prior event with a EWS, on the one hand, and an event without a EWS on the other (sometimes referred to as using 'analogues'). However, such comparisons are methodologically problematic because two climatic/weather events are never identical and many other changes to community vulnerability may have happened during the intervening period that may account, at least in part, for the changes in health outcomes. Climate-driven models which can be used to predict what would have happened in the absence of the intervention are best placed to create the counter factual for an EWS.[37]

3.7 Conclusions

This chapter highlights the need to understand the spatial-temporal scales of both the decision-context and the potentially relevant climate information. Attention has also been given to specific challenges that are associated with data issues and the identification of climate-health relationships. This and subsequent chapters highlight the need to understand the drivers of uncertainty in model development, since this understanding provides the basis for reducing it where possible. Translating research into policy and practice is a critical consideration for those engaged in developing climate services for the health sector (see § 10.3). The importance of employing a systematic approach to building an evidence-base that can influence policy cannot be over-emphasized.

Notes

i https://www.cdc.gov/mmwr/preview/mmwrhtml/mm5828a3.htm.
ii www.cochrane.org.
iii https://www.campbellcollaboration.org.
iv https://www.nice.org.uk.
v https://www.york.ac.uk/crd/.

References

1 Thomson, M. C., Ceccato, P., Lyon, B. & Macfarlane, S. B. in *The Palgrave Handbook of Global Health Data Methods for Policy and Planning* (eds. S. B. Macfarlane & C. Abouzahr) (Palgrave Macmillan, London, 2018).
2 Nissan, H., Ukawuba, I. & Thomsom, M. C. *Factoring Climate Chamge into Malaria Eradication Strategy*. 63pp (GMP WHO, Geneva, 2017).
3 Taylor-Robinson, D. C., Milton, B., Lloyd-Williams, P., O'Flaherty, M. & Capewell, S. Planning ahead in public health? A qualitative study of the time horizons used in public health decision-making. *BMC Public Health* **8**, 415, doi:10.1186/1471-2458-8-415 (2008).
4 Van Panhuis, W. *et al.* A systematic review of barriers to data sharing in public health. *BMC Public Health* **14**, 1144, doi:10.1186/1471-2458-14-1144 (2014).
5 Data Revolution Group, UN. *A World That Counts: Mobilising the Data Revolution for Sustainable Development*. 28pp (2014).
6 Paploski, I. A. D. *et al.* Time lags between exanthematous Illness attributed to Zika virus, Guillain-Barré Syndrome, and microcephaly, Salvador, Brazil. *Emerging Infectious Diseases* **22**, 1438–1444, doi:10.3201/eid2208.160496 (2016).
7 Zhao, X., Chen, F., Feng, Z., Li , X. & Zhou, X. H. The temporal lagged association between meteorological factors and malaria in 30 counties in south-west China: a multilevel distributed lag non-linear analysis. *Malaria Journal* **13**, 57, doi:10.1186/1475-2875-13-57. (2014).
8 Paaijmans, K. P., Blanford, S., Chan, B. H. K. & Thomas, M. Warmer temperatures reduce the vectorial capacity of malaria mosquitoes. *Biological Letters* **8**, 465–468, doi:10.1098/rsbl.2011.1075 (2012).
9 Mordecai, E. A. *et al.* Optimal temperature for malaria transmission is dramatically lower than previously predicted. *Ecological Letters* **16**, 22–30, doi:10.1111/ele.12015 (2013).
10 Duplantier, J. M., Duchemin, J. B., Chanteau, S. & Carniel, E. From the recent lessons of the Malagasy foci towards a global understanding of the factors involved in plague reemergence. *Veterinary Research* **36**, 437–453, doi:10.1051/vetres:2005007 (2005).
11 Eisen, R. J., Griffith, K. S., Borchert, J. N., MacMillan, K., Apangu, T., Owor, N. *et al.* Assessing human risk of exposure to plague bacteria in northwestern Uganda based on remotely sensed predictors. *American Journal of Tropical Medicine and Hygiene* **82**, 904–911 (2010).
12 Eisen, R. J., Borchert, J., Mpanga, J., Atiku, L., MacMillan, K., Boegler, K. *et al.* Flea diversity as an element for persistence of plague bacteria in an east African plague focus. *PLOS One* **7**, e35598 (2012) doi: 10.1371/journal.pone.0035598.
13 Monaghan, A. J. *et al.* A regional climatography of West Nile, Uganda, to support human plague modeling. *Journal of Applied Meteorology and Climatology* **51**, 1201–1221 (2012).
14 MacMillan, K. *et al.* Climate predictors of the spatial distribution of human plague cases in the West Nile region of Uganda. *American Journal of Tropical Medicine and Hygeine* **85**, 514–523 (2012).
15 Thomson, M. C., Mason, S. J., Phindela, T. & Connor, S. J. Use of rainfall and sea surface temperature monitoring for malaria early warning in Botswana. *American Journal of Tropical Medicine and Hygiene* **73**, 214–221 (2005).
16 Sofianopoulou, E., Pless-Mulloli, T., Rushton, S. & Diggle, P. Modeling seasonal and spatiotemporal variation: the example of respiratory prescribing. *American Journal of Epidemiology* **186**, 101–108, doi:10.1093/aje/kww246 (2017).
17 Laneri, K. *et al.* Dynamical malaria models reveal how immunity buffers effect of climate variability. *Proceedings of the Nattional Academy of Sciences* **112**, 8786–8791, doi:10.1073/pnas.1419047112 (2015).
18 Metcalf, C. J. E. *et al.* The seasonal and climatic determinants of access to care: implications for measles outbreak risk in Madagascar. *Lancet* **389** (2017).
19 Hess, J. J., Eidson, M., Tlumak, J. E., Raab, K. K. & Luber, G. An evidence-based public health approach to climate change adaptation. *Environmental Health Perspectives* **122**, 1177–1186, doi:10.1289/ehp.1307396 (2014).

20 Stanke, C., Kerac, M., Prudhomme, C., Medlock, J. & Murray, V. Health effects of drought: a systematic review of the evidence. *PLOS Currents Disasters* **5**, pii, doi:10.1371/currents.dis.7a2cee9e980f91ad7697b570bcc4b004 (2013).
21 Okoli, C. & Pawlowski, S. D. The Delphi method as a research tool: an example, design considerations and applications. I. *Information Management* **42**, 15–29, doi:10.1016/j.im.2003.11.002. (2004).
22 Evgeniou, T. How to tell if you should trust your statistical models. *Harvard Business Review: Information & Technology* (2014). Accessed at https://hbr.org/2014/09/how-to-tell-if-you-should-trust-your-statistical-models.
23 Bjørnstad, O., Finkenstadt, N. B. & Grenfell, B. T. Endemic and epidemic dynamics of measles: Estimating epidemiological scaling with a time series SIR model. *Ecological Monographs* **72**, 169–184 (2002).
24 He, D., Ionides, E. L. & King, A. A. Plug-and-play inference for disease dynamics: measles in large and small populations as a case study. *Journal of the Royal Society Interface* **7**, 271–283 (2010).
25 Laneri, K. *et al.* Forcing versus feedback: epidemic malaria and monsoon rains in northwest India. *PLOS Computational Biology* **6**, e1000898 (2010).
26 Heesterbeek, H. *et al.* Modeling infectious disease dynamics in the complex landscape of global health. *Science* **347**, aaa4339. (2015).
27 Yang, W., Karspeck, A. & Shaman, J. Comparison of filtering methods for the modeling and retrospective forecasting of influenza epidemics. *PLOS Computational Biology* **10**, e1003583. (2014).
28 Corley, C. D. *et al.* Disease prediction models and operational readiness. *PloS One* **9**, e91989, doi:10.1371/journal.pone.0091989 (2014).
29 Omumbo, J., Lyon, B., Waweru, S. M., Connor, S. & Thomson, M. C. Raised temperatures over the Kericho tea estates: revisiting the climate in the East African highlands malaria debate. *Malaria Journal* **10**, 12, doi:10.1186/1475-2875-10-12 (2011).
30 Araujo Navas, A. L., Hamm, N. A. S., Soares Magalhães, R. J. & Stein, A. Mapping soil transmitted helminths and schistosomiasis under uncertainty: a systematic review and critical appraisal of evidence. *PLoS Neglected Tropical Diseases* **10**, e0005208, doi:10.1371/journal.pntd.0005208 (2016).
31 Palmer, T. N., Doblas-Reyes, F. J., Hagedorn, R. & Weisheimer, A. Probabilistic prediction of climate using multi-model ensembles: from basics to applications. *Philosophical Transactions of the Royal Society B-Biological Sciences* **360**, 1991–1998 (2005).
32 Thomson, M. C. *et al.* Malaria early warnings based on seasonal climate forecasts from multi-model ensembles. *Nature* **439**, 576–579 (2006).
33 Ruiz D. *et al.* Multi-model ensemble (MME-2012) simulation experiments: exploring the role of long-term changes in climatic conditions in the increasing incidence of *Plasmodium falciparum* malaria in the highlands of Western Kenya. *Malaria Journal* **13**, 206, doi:10.1186/1475-2875-13-206 (2014).
34 Yamana, T. K., Kandula, S. & Shaman, J. Superensemble forecasts of dengue outbreaks. *Journal of the Royal Society Interface* **13**, pii: 20160410, doi:10.1098/rsif.2016.0410 (2016).
35 Mason, S. J. in *Climate Information for Public Health Action* (eds. M.C. Thomson & S. J. Mason), Ch. 7 (Routledge, London, 2018).
36 PHE. Heatwave plan for England. Report No. Accessed at https://www.gov.uk/government/uploads/system/uploads/attachment_data/file/429384/Heatwave_Main_Plan_2015.pdf, 45 (Public Health England, 2015).
37 Worrall, E., Connor, S. & Thomson, M. C. A model to simulate the impact of timing, coverage and transmission intensity on effectiveness of indoor residual spraying (IRS). *Tropical Medicine and International Health* **12**, 1–14 (2007).

4

CLIMATE BASICS

Simon J. Mason
Contributors: Madeleine C. Thomson

> What a strange drowsiness possesses them!
> It is the quality o' th' climate.
>
> The Tempest *by William Shakespeare*

4.1 Introduction

In this chapter, we focus on key concepts that health professionals need to understand in order to use climate data and information effectively. We explain the distinction between climate and weather and introduce some basic principles on the physics of the climate. We then go on to describe the character and behaviour of three key variables, namely temperature, precipitation and humidity. Descriptions of a range of other relevant important climate variables follow. Included is a section on storms that describes how temperature, wind and rainfall interact to create devastating extreme events. The final section discusses how data for different variables can be aggregated in time and space, and identifies challenges that emerge in the process.

4.2 What is climate?

'Climate is what you expect, but weather is what you get' is one of those well-worn quotes that is almost invariably misattributed. The credit should go to Andrew Herbertson, who was Oxford's first Professor of Geography. This adage is cleverly succinct, but, as a definition, it does not specify what 'climate' and 'weather' actually are. There may be little problem in understanding weather as how we experience the air around us, whether it is sunny or rainy, etc. (Box 4.1), but climate is too

BOX 4.1 WEATHER, CLIMATE AND CLIMES

There is an important difference between weather and climate:

- *Weather* is the state of the atmosphere as it is experienced at any moment; for example, at the time of writing in New York City on Thursday 9 March 2017, it is 14 °C and sunny. (An atmosphere is the layer of gases that surround a planet – on Earth it is the layer of air).
- *Climate* is the total experience of the weather over a period of time and in a specific location. The climate may be summarized as an average of the weather conditions; for example, last month (i.e., February 2017) the average temperature was 5 °C and there were 63 mm of rain and snow combined. However, a more comprehensive climate summary involves information about the variability and extremes of weather; for example, in New York last month the temperature ranged between –7 °C and 21 °C, there were snowfalls of between 20 and 200 mm, and the most rain on any day was 30 mm.

Scientists who study the weather will typically consider only the atmosphere, but the climate cannot be understood or predicted properly without studying how the atmosphere interacts with the oceans and the land-surface, including ice and snow cover.

The word 'climate' comes from the Greek word for 'incline', describing the curvature or slope of Earth's surface between the equator and the poles. In the 14th century 'climate' meant the area between two lines of latitude, and by about 1600, the word became associated with the weather in such zones. Latitudinal zones remain an important idea in climate (§ 5.2.2), and it is worth clarifying what these zones are because they are referred to extensively throughout this book (see Figure 4.1).

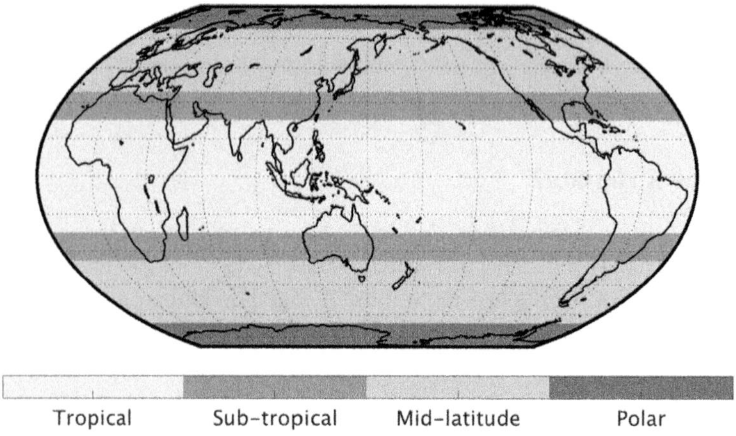

FIGURE 4.1 Latitudinal zones

- The *tropics*, sometimes called the torrid zone, lie between the equator and 23.4° latitude. Within the tropics, the sun is directly overhead at least once per year. A little over one-third of Earth's land area lies within the tropics. Most of the tropics are warm and moist all year; most areas have a distinct wet and dry season.
- The *extratropics* lie poleward of 23.4° latitude, and are subdivided into:
  - The *temperate* zone lies between 23.4° and 66.6° latitude. It has much more distinct winter and summer seasons than do the tropics. The extratropics is further subdivided into:
    - The *subtropics*, which are between 23.4° and 35° latitude. These areas usually have a warm or hot summer and a mild winter. Rainfall is seasonal: in some areas, the wet season is in the winter, in others it is in the summer. The winter rainfall areas are confusingly called Mediterranean – these Mediterranean climates occur in other parts of the world, such as California, and the south-western parts of Chile, South Africa and Australia. The summer rainfall areas are called humid subtropical.
    - The *mid-latitudes*, which lie between 35° and 66.6° latitude. Here the winters and summers are strongly affected by proximity to the sea (§ 5.2.3), and so the climates are described either as maritime (where the seasons are milder) or continental (where the seasons are more extreme).
  - The *polar* regions, which lie poleward of 66.6° latitude (some definitions use 60° latitude). Here the winter is exceptionally cold. Poleward of 66.6° latitude the sun is below the horizon (i.e., it is night-time) for a continuous period of 24 hours at least once in a year (and is above the horizon – it is day-time – for a 24-hour period at least once).

often understood as the average weather, which is too simple an interpretation. Is it possible to provide a tighter definition of climate? Unfortunately, even the American Meteorological Society's definition – 'the slowly varying aspects of the atmosphere-land-hydrosphere-system' – is technically problematic: 'slowly varying' seems to preclude weather extremes, while the 'atmosphere-land-hydrosphere-system' sounds horribly jargony.

Precise definition is essential for proper understanding, but an imprecise definition can sometimes suffice. For example, nobody knew exactly what caused cholera before the mid-19th century, but doctors knew enough long before then to quarantine the victims. In that spirit, a definition of climate adequate for the purposes of this book, but likely inadequate for the technical purist, is: 'the total experience of the weather at any place over some specific period of time'[1] (see Box 4.1), with 'some specific period of time' being understood to be at least a few years (see further discussion in § 4.3).

Even in the absence of a technical and easy-to-understand definition of climate, it is simple enough to list its important characteristics that may directly affect human health. These characteristics are discussed in the following sub-sections.

4.2.1 Temperature

Temperature is one of the most important of all climate variables for health because of its direct impact on the human body as well as many indirect impacts, such as its effects on disease transmission (see §§ 2.3 and §§ 3.4).

Temperature is much simpler to define than climate: the air temperature is how hot or cold the air is. What may be less obvious is that the air is not heated directly by the sun. Instead, with only a few minor exceptions, such as the ozone layer, the sun heats Earth's surface, and it is the surface that heats the overlying air. As a result, the atmosphere is heated predominantly from the bottom rather than from the top (while the reverse is true of the oceans). Our weather and climate would be inconceivably different if the opposite were the case, and climatologists would not be as interested as they are in Earth's surface (§§ 5.2.3 and 5.2.4). As an example, consider how land-use can have such a noticeable effect on temperature: the air temperature in parks is much lower than in open-air car parks in the summer because for many reasons vegetation does not heat up anywhere near as much as concrete and tarmac. A practical implication is that great care must be taken to measure surface air-temperature so that it is consistent with international standards: it should be measured at 2 m above the ground and from inside a shaded box (called a Stevenson screen) so that there is little chance that Earth's surface or the sun are directly heating the thermometer.

Air temperatures are commonly reported as a maximum, a minimum and/or a mean over a 24-hour period. The maximum temperature generally represents the day-time temperature and the minimum the night-time temperature, but it is possible for the day to be colder than the night, most often in the extratropics in winter (see Box 4.1 and § 5.3.1). The mean temperature is the temperature averaged over the whole day, which is not necessarily equivalent to the average of the maximum and the minimum temperature.

Temperatures are recorded most commonly as degrees centigrade (°C), or degrees Fahrenheit (°F). Fahrenheit is defined in relation to the properties of water, just as centigrade is, but with different reference values. Centigrade is defined to have a value of 0 (32 for Fahrenheit) at the freezing point of pure water at sea-level, and of 100 (212 for Fahrenheit) at its boiling point. Sea water freezes at about –2 °C, and water's boiling point decreases by about 1 °C every 300 m, so scientists have to be precise about such definitions. Celsius is based on a more accurate measurement of water's properties than is centigrade, and so, strictly, the two scales are very slightly different. Celsius is preferred in science, but it is common to treat Celsius and centigrade as synonyms. To convert temperatures in degrees Celsius to Fahrenheit, multiply Celsius by 1.8 (or 9/5) and add 32 (or, for a simpler approximation, multiply Celsius by 2 and add 30). For both Celsius and Fahrenheit the zeroes are arbitrary and so it does not make sense to calculate ratios: 5 °C is not half as cold as 10 °C.

Many scientists prefer to work in Kelvin, a scale that does have an absolute zero (but not a degree sign). If air could be cooled to 0 K (about –273 °C), it

BOX 4.2 MEASURING HOT AND COLD

Most people have an approximate sense of how hot or cold a given temperature is, but it can be helpful to have a more precise indication of what the temperature might mean in terms of its impacts (using degree days), or in terms of how hot or cold the temperature feels in the context of the current weather conditions (apparent temperatures).

Degree days

In building design and energy planning *cooling degree days* indicate the amount by which air must be cooled to reach a comfortable level for human habitation. The 'comfortable' level is location specific: in the USA, it is set to 65 °F (about 18 °C); in the UK, it is 22 °C, while in Hong Kong it is 26 °C. There are various ways of calculating the cooling degree days, the simplest of which is to count the number of degrees above that threshold. For example, if the air temperature is 25 °C, it needs to be cooled by 3 degrees (in the UK). If the temperature is 20 °C no cooling is required and so the cooling degree days are zero (cooling degree days cannot be negative). Whereas standard temperature measurements are averaged over periods longer than a day, cooling degree days are accumulated. For example, if the 25 °C day is followed by one of 27 °C, the cooling degree days are:

$$(25 - 22)\ °C + (27 - 22)\ °C = 8\ °C.$$

If temperatures drop below the threshold, the air may need to be heated to reach a comfortable level. *Heating degree days* can be calculated in the opposite way to cooling degree days. The same 65 °F threshold is used for both the heating and the cooling indices in the USA, but in the UK a heating threshold of 15.5 °C is used, and Hong Kong uses 18 °C.

In biology, an equivalent concept to cooling degree days is used to define *growing degree days (GDD),* for which it is assumed that, within limits, the growth of organisms to reach a certain development stage increases with warmer temperatures, but that there is no growth if temperature is below a threshold. Unique to each species the GDD has been extensively used in agrometeorology to predict plant and pest development, but also in medical entomology including forensic entomology as fly infestation of a corpse may provide information on the time of death.

The GDD concept also applies to the development stages of disease vectors (eggs, larvae, pupae, adult) as well as the parasites they carry (e.g., for malaria: gametocyte, ookinete, oocysts and sporozoites). Pathogens (parasitic, viral and bacterial) inside the human or vector host are at the temperature of their immediate environment (~ 37 °C in humans). However, as insects and tick vectors do not regulate their temperature the internal temperature (where pathogens develop) approximates that of the vectors' immediate external environment. Changes in environmental temperature may dramatically affect the development rate of both the vector and the pathogen(s) it contains.

Apparent temperatures

Apparent temperatures measure how hot or cold the air feels: a hot day feels even hotter if it is muggy, and a cold day even colder if it is windy. *Heat indices* and *wind chill factors* are examples of apparent temperatures that account for these different perceptions.

Various heat indices have been devised to combine temperature and relative humidity, the details of which vary from country to country. These indices are often used as heat wave metrics, and are designed to give an indication of the apparent temperature. Even the simplest and most commonly used heat indices have complicated formulae, but all involve exponential increases in apparent temperature as the humidity rises (i.e., the apparent temperature increases more quickly the closer the relative humidity approaches 100%), and the rates of increase are faster the higher the temperature. However, few heat indices account for exposure to the sun or for how windy it is, and they make simple assumptions about clothing and body-type. Therefore, such indices should only be used as approximations. For many practical purposes, a heat index may underestimate the apparent temperature, and possibly quite substantially.

A breeze can have a cooling effect, which is why fans can be effective even in the absence of air conditioning. The cooling effect occurs partly because wind increases the evaporation rate so that sweating can cool us down more effectively, and because the air that is warmed by our body is quickly replaced by colder air. However, if the air temperature is above body temperature, the apparent temperature will increase because the air is now cooled rather than warmed by our colder body and that cooled air will be quickly replaced by hotter air. Appropriate clothing in very hot weather can reduce that effect. Dark loose clothes are worn by the Touregs of the Sahel and the Bedouins of North Africa and the Middle East in very hot and dry environments. While the black cloth absorbs heat from the sun more readily than white cloth, the loose dark clothes encourage convection, releasing warm air as it rises away from the body and bringing in cooler air to replace it.

While only a few of the more sophisticated heat indices account for the effects of wind in warm and hot weather, cold indices do factor in the chilling effect of wind. Wind has a chilling effect regardless of whether the air is hot or cold: it rapidly evaporates any moisture on our skin (which is why it is so important to stay dry in the cold), and quickly replaces the air immediately in contact with our body. *Wind chill* factors are often calculated for winter temperatures to account for such effects. The air temperature is measured, and then a complicated formula is applied to subtract a wind chill depending on the wind speed. The wind chill temperature is always less than the actual temperature unless the air is calm, in which case the two temperatures are equal.

would contain no heat energy at all. The Kelvin scale is not used widely for public communication. Instead, common use is made of various temperature indices that attempt to indicate our subjective sense of how hot or cold the air is, taking into account the humidity (§ 4.2.3) and/or wind chill (§ 4.2.4) and possibly other concurrent weather conditions (Box 4.2).

Temperature has an important effect on air that is critical in producing our weather. Hot air expands and becomes less dense, and will be forced to rise by the surrounding cooler and denser air. It is because of this buoyancy of hot air that we warm our hands above a heater, and do our baking on the top shelf of the oven where the temperature is highest. In contrast, cold air contracts and is dense, which is why cold air seeps into a room at the bottom of the door. This effect of temperature on the buoyancy of air is critically important in the formation of rainfall (§ 4.2.2) and wind (§ 4.2.4) and in the development of many types of storm (§ 4.2.8).

There is an important distinction between temperature and heat. To melt ice you must apply heat to increase its temperature to about 0 °C, but then you must continue to apply heat to convert the ice into water without actually increasing the temperature. The amount of energy required for the melting is considerable, which is why ice cubes act to cool drinks so effectively. Heat in the beverage turns the solid ice into liquid water – the temperature of the drink cools down to melt the ice, but although the ice is melting, it does not get any warmer. It takes the same amount of energy just to convert ice at 0 °C to liquid water at 0 °C as it does to heat water from 0 °C to about 80 °C. For this reason, if you mix ice at 0 °C with the equivalent amount of water at boiling point, the temperature of the melted mix will be only about 10 °C rather than 50 °C because most of the heat of the boiling water is used for the melting.

This same effect occurs when boiling water, but even more heat is required to convert water to vapour (as distinct from steam, which is small drops of liquid water, exactly like a cloud; water vapour is an invisible gas) than it is to convert ice to water. To convert water at 100 °C to vapour at 100 °C takes more than five times the energy that is needed to heat water from 0 °C to 100 °C. The heat that results in a temperature change is called *sensible* heat because it can be felt; heat that converts solid (ice) to liquid (water) or liquid to gas (water vapour) is called *latent* heat. When water vapour condenses, or when water freezes, the latent heat is released and can result in sensible heating of the air. Because water vapour contains so much latent heat, vapour in the air is exceptionally important in climate (§ 4.2.3), in addition to its properties as a greenhouse gas (§ 5.4.2.3).

4.2.2 Precipitation

Rain, snow, hail and sleet (a mix of rain and snow) are the main forms in which water falls to Earth's surface as either liquid or solid. Collectively, these forms are called 'precipitation'. In most parts of the world, rain is the most dominant form, and so in this book 'rainfall' is used to refer to all forms of precipitation unless the context makes clear otherwise. Water, largely obtained from rainfall, is vital to every living organism; it provides the water that plants and animals, including humans, need to thrive. In addition, there are some immediate and obvious hazards posed by the physical properties of falling hail, and of rain and snow on the ground.

For rainfall to occur, water vapour in the air must somehow condense to form clouds, and then the droplets must coalesce to form ice or water drops heavy enough to fall to Earth's surface. The condensation of the water vapour requires the air to cool, and cooling can occur for different reasons. The most important cooling process is for air to rise (see further discussion in §§ 4.2.3 and 5.2.1). Different types of rainfall occur depending on how air rises:

1. *Convective* rainfall occurs because air is heated (e.g., by passing over a warm surface such as the equatorial ocean) and therefore becomes more buoyant than surrounding areas (or the same air cools less quickly than its surroundings)
2. *Large-scale* rainfall occurs when warmer air meets colder air: the warmer air rises above the colder air, or the colder air undercuts the warmer, because of differences in buoyancy
3. *Orographic* rainfall occurs when air is blown and forced over a physical obstacle such as a mountain

These different types of rainfall vary in their relative importance in space and time (see § 5.2 and 5.3). Although more than one of these cooling processes can occur at the same time, and although there are many exceptions, the different types of rainfall can often be distinguished by how they are experienced. Convective rainfall is typically intense, localized and short-lived, and is often associated with thunderstorms. Large-scale rainfall is often light, widespread and may last many hours or even a few days. Orographic rainfall is less easily characterized, but is often an enhancement of convective and/or large-scale rainfall; it is experienced when you ascend a mountain and it starts raining, or becomes damp because you have entered the clouds. With few exceptions, those clouds have likely formed because of the orographic effect.

Daily precipitation is measured as an accumulation in millimetres over a 24-hour period (the starting time of day varies by location). Meteorologists measure precipitation as the depth of water (snow and hail are first melted to get an equivalent amount of rainfall) that has fallen to the surface. However, for public communication, snow is usually measured by its depth on the ground, and is generally between two and 20 times as deep as the equivalent in rainfall. The most common method of measuring precipitation is to capture it in a specially designed bucket, known as a rain gauge. Even with the most well-designed rain gauges, it is very difficult to measure precipitation accurately, primarily because the gauges disrupt small-scale wind patterns in their immediate vicinity, so that less falls into the gauge than onto the ground nearby. Gauges typically underestimate the precipitation that does fall; for snow, this underestimation may be as much as 35% or more.[2]

Many climate models measure precipitation as gridbox-averaged values in contrast to the point-specific measurements of rain gauges, and therefore typically use units of metres per second (or, equivalently, as kilograms per square metre per second). This convention is strictly in accordance with SI units, but it does yield ridiculously small numbers: 25 mm of rain is a reasonably wet day for many places, but converts to 2.9×10^{-7} m.sec^{-1}.

4.2.3 Humidity

Humidity is a measure of the amount of water vapour in the air. If humidity is high, evaporation slows and our body's ability to regulate its own temperature is compromised. If humidity is low, evaporation accelerates and creatures that thrive in moist environments, like mosquitoes and other disease-carrying insects or ticks (vectors), may not be able to survive (Case Study 4.1). However, humidity is only a major concern to humans in hot conditions, partly because of the biological effects of humidity in high temperatures. Sweating becomes ineffective in muggy air, but the air only feels muggy when temperatures are high. Therefore, how we experience humidity depends on temperature, and even the simplest indices of how hot the air feels involve complex adjustments for humidity (Box 4.2).

Relative humidity is the most commonly reported measure of humidity in public weather information because it gives a good indication of the likelihood of rain or fog. It measures the amount of water vapour in the air as a fraction of the amount in 'saturated' air of the same temperature. Air is saturated when humidity is increased to a point beyond which the water vapour condenses more quickly than it can evaporate. The amount of water vapour required to saturate air increases by about 7% for every 1 °C warming in air temperature, which means that humidity works like compound interest: the maximum humidity increases faster and faster the hotter the air gets. As a result, at high temperatures small changes in temperature can have a dramatic effect on humidity even if the amount of moisture in the air stays the same. This sensitivity of humidity at high temperatures has many important climate implications (§ 5.2.1), but it also means that there is a need to recognize that some measures of humidity may indicate changes in temperature rather than changes in the amount of moisture in the atmosphere.

CASE STUDY 4.1 DISPERSAL OF PATHOGENS AND INSECT VECTORS – THE IMPORTANCE OF HUMIDITY AND WIND

Madeleine C. Thomson, IRI, Columbia University, New York, USA

Pathogenic micro-organisms such as viruses, bacteria and fungal spores may disperse over varying distances by air currents; ultimately coming into contact with individuals and causing disease. Short distance dispersal via tiny droplets suspended in the air (e.g., from coughing) occurs when the size of the pathogenic particle at origin is large (> 5μm) and particle desiccation is low (i.e., in a humid environment). In these circumstances, the weight of the droplet means that it readily falls to the ground and transmission is only possible when susceptible individuals are physically close to the source of infection.

In contrast, airborne transmission occurs over considerable distances when pathogenic particles are comparatively small (< 5μm) and thus can remain suspended in air for long periods (e.g., more than a week). Desiccation of large airborne droplets may rapidly reduce larger particles to below this threshold and ensure that these droplets are also aerosolized. Figure 4.2 illustrates the relationship of concentrations of bacteria in the air and relative humidities in towns and villages in Mali.[3] The winter flu season in temperate regions of the world is driven, in part, by the low absolute humidities that occur during this period.[4]

Air-borne dispersal of insect vectors is a significant component of their biology and is highly dependent on wind direction, speed and turbulence.[5] Dispersal may be passive through downwind movement of the flies that then encounter new habitats, or upwind in response to chemicals such as CO_2 that may signal the presence of a potential human or other animal host.

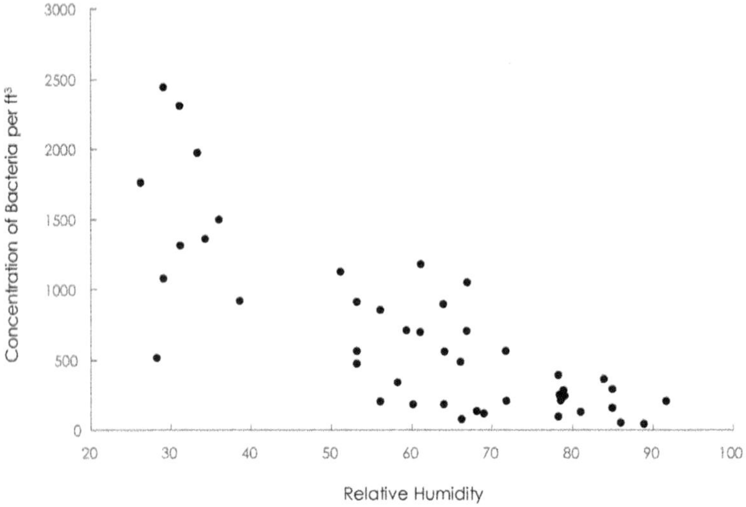

FIGURE 4.2 Concentration of airborne bacteria in relation to humidity in Mali

Long distant air-borne dispersal of insects can be very important in vector-borne disease control programmes as they can create problems of recolonization of areas where vector populations have been eliminated. Dispersal may happen over long distances when insects are caught up in local air turbulence, and then transferred to the upper atmosphere. Because air turbulence is usually greatest during the day, blackflies and day-flying mosquitoes (such as *Aedes* spp.) are more likely to be swept into the upper air and travel hundreds of miles than mosquito species that are active at night (such as the malaria carrying *Anopheles* spp).

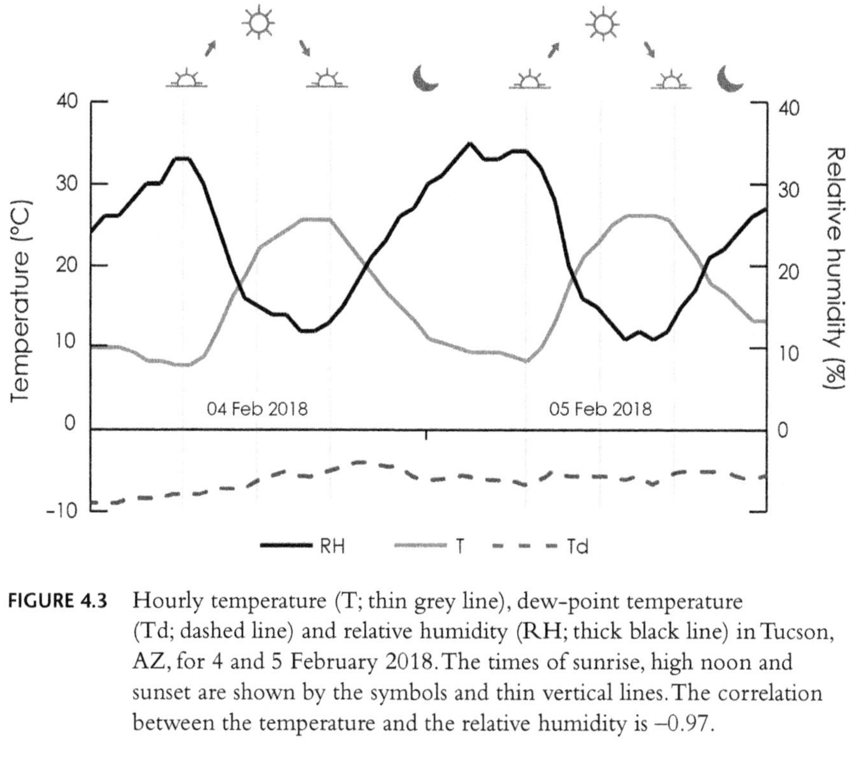

FIGURE 4.3 Hourly temperature (T; thin grey line), dew-point temperature (Td; dashed line) and relative humidity (RH; thick black line) in Tucson, AZ, for 4 and 5 February 2018. The times of sunrise, high noon and sunset are shown by the symbols and thin vertical lines. The correlation between the temperature and the relative humidity is –0.97.

Only some measures of humidity are sensitive to temperature; those that are sensitive can be affected significantly by the time of day the measurement is taken. For example, Figure 4.3 indicates how relative humidity varies inversely with the air temperature while the dew-point temperature (which is a better measure of how much moisture there is in the air; see Box 4.3) remains fairly constant. The relative humidity drops as the temperature rises simply because at higher temperatures saturated air contains more moisture. Care therefore needs to be taken to use a humidity measure that is appropriate for analyses of health effects. In fact, some measures of humidity, including relative humidity, may even be highly misleading in analyses of health impacts unless their co-variability with temperature is accounted for adequately.

BOX 4.3 WHICH IS THE MOST USEFUL HUMIDITY METRIC?

The effects of humidity depend not only on how much water vapour is in the air, but also on the temperature of the air. For example, the air does not feel muggy when it is cold, regardless of the humidity. When analysing the health effects of humidity it is therefore important to use a metric that accounts for air temperature adequately. In dynamical disease models, the compounding effects of temperature should be accounted for; in which case the pitfalls mentioned below can be ignored. However, in empirical (statistical) models, using an inappropriate metric can lead to highly misleading results. Common choices are:

- *Absolute humidity*: the mass of water vapour (in grams) per cubic metre of air. Sometimes the mass is expressed as a component of the total air pressure (§ 4.2.7.1); this component is called the vapour pressure. Because air expands as its temperature increases, a cubic metre contains more air if it is cold than if it is hot; for a fixed absolute humidity the amount of water vapour is the same. Absolute humidity is likely to be unsuitable for most analyses of health effects.
- *Specific humidity*: the mass of water vapour for a given mass of the air rather than a given volume, thus correcting for the problem with absolute humidity. Evaporation rate is directly related to the specific humidity.
- *Relative humidity*: the total amount of water vapour as a fraction of the amount in 'saturated' air of the same temperature. The amount of water vapour in saturated air increases exponentially with temperature, and so the relative humidity can show a decrease as the day heats up, and an increase again during the night (Figure 4.3). The time of day at which the relative humidity is taken must therefore be noted carefully; a daytime value may be too dependent on the temperatures, and daily average relative humidity may be quite meaningless. Relative humidity is likely to be a misleading variable in analyses of health effects.

Although relative humidity can be misleading because of its sensitivity to temperature, it does have the merit of comparing the humidity with the humidity of saturated air. Instead of subsuming this difference in a ratio, it can be expressed as a difference (or *saturation deficit*) in absolute or specific humidity or vapour pressure.

There are also indirect ways of measuring humidity that may be useful in analyses of health effects:

- *Dew-point temperature*: the temperature at which air would become saturated if it were cooled. The dew-point temperature indicates at what temperature fog or precipitation would start to form if the air cooled down.

- *Wet-bulb temperature*: the temperature to which the air would cool by evaporating the additional moisture needed to reach saturation point. That may sound rather complicated, but the idea is similar to the dew-point temperature. One virtue of the wet-bulb temperature is that it is relatively easy to measure, and is widely measured and consequently commonly used in studies of heat stress. The other humidity measures are often calculated from the wet-bulb temperature (along with the regular temperature and air pressure).

4.2.4 Wind

Wind is the flow of air, measured in terms of its direction and speed. Watch a leaf blow in the wind, and it is immediately obvious that wind speed and direction are highly variable over very short distances and times. To minimize the effects of these vagaries, the standard is to measure wind speed and direction 10 m above ground, where winds are steadier, and averaged over a two-minute period.

The direction of the wind is the direction from which it comes – a northerly wind is a wind coming from the north, blowing towards the south. Similarly, a sea-breeze is one that blows from the sea onto the land. To help in remembering that convention, think of the wind as a traveller – a German tourist is a tourist from Germany not a foreign tourist going to Germany. (Ocean currents, on the other hand, are more like pilgrims, being named for the direction in which they are going; so a north*ward* – as distinct from a north*erly* – current is one flowing towards the north.) Of course, air can also move up or down, and meteorologists have a variety of pretentious-sounding terms for these motions depending on why the vertical motion is occurring. These vertical motions are important in the formation of clouds and rain, for example, and can be important for the long-range dispersal of pathogens and insect vectors (Case Studies 4.1 and 4.2).

The prevailing winds indicate the direction from which wind comes most frequently. The dominant winds are the direction of the strongest, rather than the most frequent, winds, although they are often the same. The prevailing and dominant winds can be identified from a history of wind speeds and directions. In much of the tropics (Box 4.1), the prevailing winds are the Trade Winds, which are easterlies (i.e., blowing from east to west); in the mid-latitudes, westerly winds prevail. This contrast in wind direction between the tropics and the mid-latitudes is an effect of Earth's rotation. Monsoons comprise a major wind system that seasonally reverse their direction. The most prominent monsoons occur in tropical regions in South Asia, Africa, Australia and the Pacific coast of Central America (Figure 4.4). The prevailing winds change direction in the different seasons because of contrasts in how the land and the sea heat up in summer and cool down in winter (§ 5.2.3). The change in wind direction is often associated with a transition from a wet to a dry season, or vice versa. As a result, monsoon areas are often identified by their rainfall seasonality rather than by the winds per se (as in Figure 4.4). In the mid-latitudes,

CASE STUDY 4.2 THE WIND IN THE WEST: HOW THE ONCHOCERCIASIS CONTROL PROGRAMME FOLLOWED THE INVASION OF BLACKFLY VECTORS FROM THE SAHEL TO SIERRA LEONE

Madeleine C. Thomson, IRI, Columbia University, New York, USA

Prior to the 1970s, onchocerciasis (known as river blindness) was largely ignored by the international health community. Its devastating effects were borne by rural populations of West Africa living near the fast-flowing rivers of the Sahel such as the Volta and Niger River. When the Onchocerciasis Control Programme (OCP) was started in 1974, with the support of the World Bank and other donors, some of West Africa's richest river lands were uninhabited. In villages sited in river valleys near the breeding sites of the vector of onchocerciasis, the blackfly (*Simulium damnosum s.l.*), it was not unusual to find 60% of the adults afflicted with this filarial disease and 3% to 5% blind. Communities were forced to abandon their villages en masse to avoid this devastating disease.

Today, more than 40 years after the programme was first launched, the disease has been controlled through one of the most successful public health campaigns in history.[6] The OCP was initially based on vector control: routine spraying of identified vector breeding sites being the primary control method. In 1989 mass treatment of populations exposed to onchocerciasis with Mectizan (ivermectin) began in the OCP region and became an important component of the control strategy, both as a complement to larviciding in specific areas and as the sole intervention in most of the 'extension' areas of the OCP.[7]

Simulium damnosum s.l. species complex comprises many distinct sibling species with varying capacities to transmit the filarial worm, *Onchocerca volvulus*. Understanding the ecology of the vectors of onchocerciasis has been key to their successful control.[8] The distribution of the different members of the *S. damnosum s.l.* species complex is generally related to phtyogeographic zones (e.g., forest and savannah) but seasonal changes in their distribution occur on an annual cycle with the monsoon/harmattan winds of West Africa, which bring the rains. The winds aid dispersal of these day-flying flies when they are swept up into the upper air, where they travel average distances of 15–20 km daily and may migrate over a total distance of 400–500 km.[9] Unusual migrations of savannah species of *S. damnosum s.l.* (the species most commonly associated with the blinding form of the disease) into the forest zones of West Africa have been observed,[10] leading to speculation on the possible role of deforestation and rainfall decline on the distribution of different species of the disease.[11,12]

The rains also result in enhanced river flow and the creation of the white-water rapids that are the favoured breeding sites of these vectors. From the start of the OCP, scientists sought to understand the region's weather systems in order to track the movement of the flies, and the organization invested in hydrological monitoring of the region's river systems in order to identify where and when breeding might occur and the type of insecticide required to treat specific areas.[13]

While vector control successfully interrupted the transmission of the *Onchocerca* parasite in many areas the introduction of the anti-filarial drug Mectizan (ivermectin) led to the rapid decline in morbidity associated with the disease.[7]

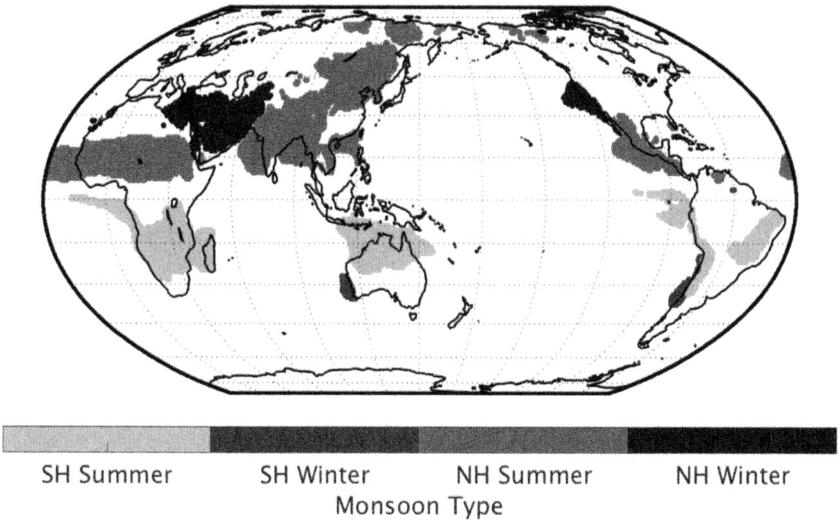

FIGURE 4.4 Locations of monsoon regions, as defined by areas that receive at least 70% of their annual rainfall during May–September (Southern Hemisphere [SH] winter / Northern Hemisphere [NH] summer) or November–March (SH summer / NH winter). The monsoons are labelled by the season in which the winds bring rain. (Data source: ECMWF Interim Reanalysis, for January 1981–March 2010[14])

changes in wind direction can have a very important effect on the air temperature at weather timescales (as discussed in § 5.2.5.1).

Wind speeds are highly variable, and so a distinction is sometimes made between gusts and sustained winds. Sustained wind speeds are calculated over two-minute periods, while gusts are instantaneous speeds and are recorded only if they exceed the sustained winds by more than 10 knots (19 $km.h^{-1}$). Knots are often used to measure wind speeds at sea (a knot is one nautical mile per hour, and a nautical mile is about 15% longer than a mile, i.e., about 1.85 km), but the standard unit of measurement is metres per second. Wind speeds are most easily interpreted when they are compared to a specially designed scale. The Beaufort scale is one such scale that relates different wind speeds to their visual effects over land or sea, and distinguishes between calm, breezes, gales and storm-force winds. There are similar scales for winds associated with tornadoes (the Fujita scale), hurricanes (Saffir-Simpson scale) and other storms (§ 4.2.8). Just as the technical names for tropical storms differ ('hurricanes' for the North Atlantic and Northeast Pacific Oceans, 'cyclones' for the Indian and South Pacific Oceans, and 'typhoons' for the Northwest Pacific) so do the scales. To add to the inconsistency, the period for measuring the sustained winds also varies. There are no equivalent scales for winter storms, which predominantly affect the mid- and polar latitudes.

Strong winds have an obvious health concern because of the physical hazards they bring through flying and falling objects. The potential for winds to cause damage increases four times for every doubling in the wind speed, but the risk is exacerbated when there is a combination of wind and rain or snow. Hence, strong tropical or winter storms can be particularly devastating (§ 4.2.8). However, wind can be a problem independently of these physical hazards: winds affect our perception of temperature and humidity because of their effect on exposure and on the evaporation rate (Box 4.2), and they transport and disperse dust, pollution, pathogens and insect vectors (Case Studies 4.1 and 4.2). This dispersal effect by the prevailing winds helps to explain land-use patterns in many cities: in mid-latitude cities, for example, the wealthier areas are often located on the western side, upstream of urban pollution sources.[15]

4.2.5 Solar radiation

As discussed above, air temperature is measured in the shade. Stand in the sun on a summer's day and it can feel extremely hot, or stand in the sun on a calm, cold winter day, and it can feel pleasantly warm. However, the air is not actually warmer in the sun; it only feels hotter in the sun because the sun is heating you directly, just as it heats Earth's surface. The most sophisticated heat indices take this exposure to the sun into account, but most indices ignore this effect and so apparent temperature values in the sun may be considerably higher than those reported or forecast for the shade (Box 4.2).

A simple measurement of the amount of sunshine is a count of the number of hours of direct sunlight. That is a rather simplistic measure for health purposes since it does not take into account the intensity of the sunlight. A more useful measurement is that of the strength of solar radiation at different wavelengths (e.g., ultraviolet, visible light, infrared); the exact units depend on how the wavelengths of interest are defined, but it is sufficient for our purposes to note that the brightness of a lightbulb (Watts) is measured in a similar way.

Of greatest concern is likely to be the amount of ultraviolet radiation reaching the surface because of its association with skin cancer and plant damage. Ultraviolet radiation also increases the formation of near-surface ozone, which causes respiratory problems and reduces the rate of photosynthesis (see Box 4.4 and § 4.2.6).

BOX 4.4 OZONE

One form of air pollution that often causes confusion is ozone. Most ozone is formed by the action of the sun's ultraviolet radiation on oxygen molecules above about 10 km. Ozone is created primarily at tropical latitudes, but large-scale wind patterns at these high altitudes move ozone toward the poles, where its concentration builds up. Ozone molecules high in the atmosphere are good for human health because they absorb harmful solar radiation. However, when ozone is found near Earth's surface (where high levels occur in association with smog), it is harmful to animal, including human, respiration as well as plant growth.

Since the 1970s, there has been concern about the 'ozone hole', which is a region of exceptionally depleted ozone over Antarctica.[16] The hole was caused by the presence of chlorofluorocarbons (CFCs) and halons – gases formerly used in aerosol spray cans and refrigerants. The hole, which is better described as a thinning, occurs primarily during the Antarctic late-winter when changes in wind patterns prevent ozone from being blown into the region from other areas. These natural wind patterns in the winter are a rather complicated result of Earth's rotation and the fact that Antarctica is very cold (§ 5.2.2.2). The hole is weak or absent over the Arctic because the Arctic is much warmer than the Antarctic.

Concerns about the effects of the ozone hole prompted a highly successful international agreement to curb gases that destroy ozone. This agreement, known as the Montreal Protocol, was signed in 1987, and resulted in the rapid phase-out of most of the relevant pollutants. After reaching its maximum size in 2006, the ozone layer is slowly being restored, but because the pollutants can remain in the air for decades, full restoration will only be achieved mid-century at the earliest.

4.2.6 Air quality

Another problematic American Meteorological Society definition is that for air pollutants: 'substances that do not occur naturally in the atmosphere'. It is not hard to think of substances (such as plates thrown in anger or children's kites) that might only qualify as pollutants from rather odd (but possibly valid) perspectives. Most people would consider an air pollutant to be a harmful substance (which does not preclude thrown stones, but might preclude kites). Again, without going into the intricacies of reaching a technically exact definition, it seems reasonable to assume that an air pollutant will be a gas, a liquid droplet or a solid no bigger than a small particle (thus precluding thrown plates) so that the pollutant can potentially stay in the air for at least a while.

The focus of the Society's definition on unnatural substances is also unsuitable for our purposes. Volcanoes, a natural phenomenon, can be an important source of toxic gases and particles that pose a health hazard. In addition, many gases that occur naturally at low concentrations may only be considered a pollutant at higher concentrations – for example, the increased greenhouse gas concentrations resulting from human activity (see § 4.2.7.3 on air chemistry, and discussions on greenhouse gases in Chapters 5 and 9).

Air pollution is a major cause of death globally.[17] It affects humans directly by causing a wide range of disease and allergies, and indirectly through infectious disease and nutrition by weakening people's immune systems, by facilitating the transmission of airborne diseases, by affecting the productivity of crops and livestock, and by toxifying water and food supplies. Pollution levels are well-correlated with other meteorological variables: levels can peak during some types of cold weather conditions and during heat waves, for example. Additionally, rainfall typically cleans the air of many pollutants.

4.2.7 Other important meteorological parameters

Meteorologists study additional parameters in order to understand the weather and climate more precisely, and to make forecasts. Some of these parameters are also important for human health and include:

4.2.7.1 Air pressure

Air pressure (sometimes called barometric pressure) is a measure of the weight of the air above a given point. Air pressure is of interest to meteorologists because spatial differences in air pressure contribute to the formation of wind. Air pressure varies horizontally by only a few percent, but it decreases quickly with altitude: the amount of air decreases by about 11% for every kilometre in altitude (§ 5.2.1.1). Because of this sensitivity to altitude, surface pressure is adjusted to sea-level pressure (these adjustments can become unrealistic from high-altitude locations). The rapid decrease of air pressure with altitude is why it is difficult to climb high mountains, for example, and is one reason why airplane cabins have to be pressurized. At high altitudes, altitude sickness, caused by the lack of oxygen, can occur. For example, in La Paz, Bolivia, the world's highest capital city (about 3650 m), there is about 40% less air, and therefore less oxygen per breath, than at sea level. The city was temporarily barred from hosting international soccer matches because of the difficulty visiting teams experienced in acclimatizing to the lack of oxygen. At more moderate altitudes, where humidity tends to be low, many people find the air more comfortable than they do at sea-level.

4.2.7.2 Geopotential heights

Air pressure aloft explains wind patterns higher up in the atmosphere. Rather than measuring air pressure at different altitudes, it is more useful (to meteorologists) to measure at what height a particular air pressure is observed. This height is measured in geopotential metres, which are slightly different from standard metres because they take into account minor differences in the strength of gravity (which strengthens slightly towards the equator, for example). Because air contracts as it cools (which is why a sealed, half-filled plastic bottle of water crumples when it is left in the freezer), pressure decreases more rapidly with altitude if the air is cold than if it is warm. As a result, small differences in pressure at or near the surface can become enhanced or reversed higher up if there are differences in temperature. These effects are responsible for the formation of the important jet streams (see § 5.3.2).

4.2.7.3 Air chemistry

The chemical composition of air is important because different gases have different properties. Most gases, with the notable exception of ozone, do not absorb the sun's radiation (shortwave radiation). Instead, they absorb the heat emitted from Earth (longwave radiation), which is why the atmosphere is heated from Earth's surface rather than by the sun (§ 4.2.1). However, some gases are more effective at absorbing heat emitted from the surface than others are, thereby contributing disparately to the temperature of the air (see the discussion on the greenhouse effect in Chapters 5 and 9).

4.2.7.4 Sea, land and ice

Because Earth's surface is the primary source of heat for the air, meteorologists closely monitor sea and land-surface temperatures (see Chapters 5 and 7). High land temperatures can lead to heat waves: land temperatures increase the most when the soil is dry because less energy is required to heat up land than water. Hence, soil moisture is an important influence on air temperature. Sea-surface temperatures, however, are the most important surface variable, partly because 70% of Earth's surface is sea, but also because the sea is the main source of atmospheric moisture that eventually falls as rain (as well as providing most of the energy for the atmosphere as discussed below). Sea-surface temperatures are measured by ships, specially deployed buoys and other instruments, and by satellites (see § 6.2.1). Measurements of sub-surface temperatures, ocean currents, salinity and sea-level height are required to understand how sea-surface temperatures change over time (Chapters 8 and 9).

Snow and sea-ice depth and extent are important parameters: snow and ice reflect much of the incoming sunlight, and as a result limit the warming experienced over a snow- or ice-covered surface. Because the ice and snow are cold, the overlying air may also cool and more snow and ice may form, causing further cooling. This cycle also works in the opposite direction: if snow melts, the uncovered surface can be heated from the sun, and it reflects less of the sunshine so there is stronger heating and further melting. The cycle is called the albedo effect – albedo is a measure of the proportion of sunlight that is reflected.

4.2.8 Hurricanes, typhoons and other storms

Often we may be more concerned with a combination of weather conditions rather than with the individual components described in the previous sub-sections. Some of the biggest impacts are associated with storms, which are examples of extreme weather conditions involving strong winds and/or heavy rain or snow (§ 4.2.2), and possibly thunder and lightning. The combination of strong winds and heavy rainfall or snow can be highly hazardous, increasing the chances of crush injuries from trees falling and buildings being damaged, for example. There are many different types of storms, depending on their size, cause and what type of precipitation there is, if any.

The largest storms are cyclones, which are areas of low air pressure (§ 4.2.7.1) around which winds blow. Tornadoes are much smaller systems, and are described in § 4.2.8.3. The winds blow around a cyclone in the same direction as Earth is spinning (anticlockwise in the Northern Hemisphere, and clockwise in the Southern Hemisphere). Although we may not physically notice it, the low pressure is an important characteristic of a cyclone: low pressure causes winds to converge into the cyclone, bringing moisture from surrounding areas that can fall as rain. The water vapour that the winds bring to the storm also provides the main fuel to maintain the storm (as explained below).

For a cyclone to last more than a few minutes, the low pressure at the centre needs to be maintained, despite the winds bringing air into the cyclone to increase the air pressure. There are two important ways in which the central low pressure can persist or even strengthen:

1. Winds may be deflected as they approach the centre as an effect of Earth's rotation (the *Coriolis effect*). When an area of low pressure forms close to the equator, the winds can blow directly into that area so that air pressure differences smooth out quickly. Beyond about 5° of latitude, Earth's rotation deflects those winds and so a cyclone can persist and develop. Away from the equator, it is thus possible for cyclones to form.
2. Air near the centre of a low-pressure region may rise. Differences in the reasons for the rising air are how scientists distinguish between tropical and extra-tropical cyclones.

4.2.8.1 Tropical cyclones

Tropical cyclones form between about 5° and 30° latitude (Figure 4.5). The strongest tropical cyclones in the North Atlantic are called hurricanes, and those in the Northwest Pacific are called typhoons. Elsewhere they are called tropical cyclones, but this name can be used generically for all regions.

The low pressure in tropical cyclones is caused in part by strong surface heating, which, in turn, causes the overlying air to expand and rise because of its buoyancy and latent heat release (§§ 4.2.1 and 4.2.3). If the air can keep rising, the air pressure

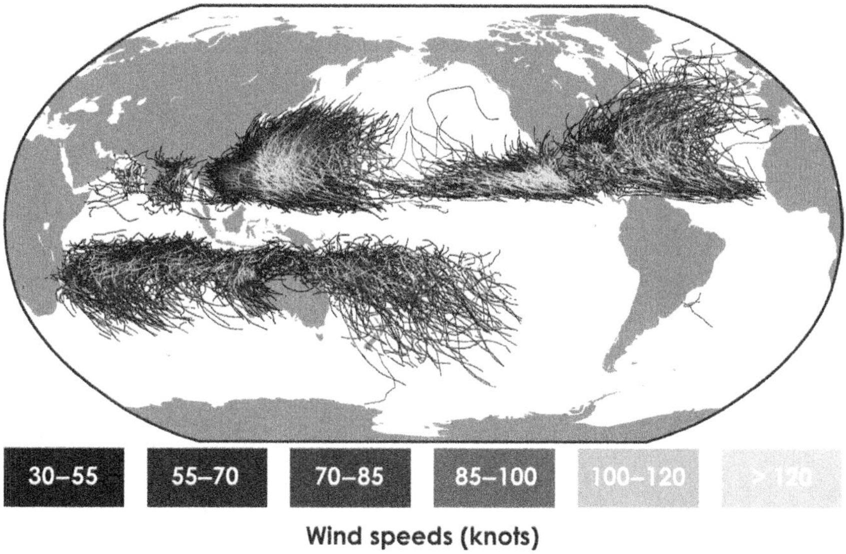

FIGURE 4. 5 Global distribution of tropical cyclone tracks, 1991–2010 (Data source: International Best Track Archive for Climate Stewardship (IBTrACS) version 3)

in the centre of the cyclone will stay low. Because tropical cyclones rely on warm humid air, they develop only over very warm ocean surfaces. More specifically, tropical cyclones that reach the strength of hurricanes and typhoons form only where the sea-surface is warmer than 26.5 °C. Even this threshold temperature is insufficient by itself: these high sea temperatures need to extend to at least about 50 m below the sea surface because otherwise the strong winds would cause waves that would mix the surface of the ocean with colder layers below, which would weaken the cyclone.

Because strong tropical cyclones require so much latent energy, they form only in the summer and autumn when the sea is warmest, and quickly dissipate when they move over land where their source of moisture is cut off. The importance of a warm ocean explains why the largest and strongest tropical cyclones are the western Pacific typhoons – the warmest oceans are in the western tropical Pacific (see Box 5.1 on El Niño).

The damage caused by a tropical cyclone is closely related to its strength because the destructive force of winds increases four times for each doubling of the wind speed (§ 4.2.4), so that even small increases in wind speed can cause considerable extra damage. The location of the strongest winds in a tropical cyclone depends on the direction the cyclone is moving and where the cyclone is: if you face the direction the storm is heading the strongest winds are on the right side of the eye in the Northern Hemisphere (in the Southern Hemisphere they are on the left side). They are strongest here because the wind blowing around the cyclone is in the same direction as the movement of the storm and so the wind speed is the speed of the wind circulating around the low pressure plus the speed of movement of the cyclone.

Just as wind speed increases with the strength of a tropical cyclone, so does the intensity of rainfall, but the speed of movement of the storm is often a more significant determinant of damage than the rainfall intensity. If the storm is moving slowly, heavy rainfall will occur for a long time, and so the severest flooding occurs in slow-moving tropical cyclones rather than in those that are more intense and faster-moving. Of course, the severest flooding occurs when the tropical cyclone is both slow-moving and intense. Examples include: Hurricane Mitch in 1998, which caused devastating floods in Honduras, Guatemala and Nicaragua; Hurricane Harvey in 2017, which caused extensive flooding in Texas; Typhoon Koppu in 2015, which flooded the northern Philippines; and Cyclone Sidr in 2007, which resulted in one of Bangladesh's worst natural disasters.

4.2.8.2 Extratropical cyclones

In tropical cyclones, the low pressure is maintained by rising air that is warm and moist; in extratropical cyclones, the rising air is caused by the jet streams many kilometres up. In effect, therefore, tropical cyclones are forced from the surface by strong heating, whereas extratropical cyclones are forced by strong winds high above Earth's surface. The mechanisms involved are complicated, but extratropical cyclones form along the boundary between warm and cold air, which also contributes to the formation of the jet streams. This boundary between warm and cold air is called a 'front'.

Like tropical cyclones, extratropical cyclones can bring heavy rain and strong winds, but the pattern of rain is different to that in a tropical cyclone. In a tropical cyclone, heavy rain occurs near the eye where the air is rising fastest; in an extratropical cyclone heavy rainfall occurs where there is large-scale rising of air along the boundary of the warm and cold air (§ 5.2.5.1). Thus, extratropical cyclones are responsible for the 'large-scale' rainfall described in § 4.2.2.

Unlike tropical cyclones, extratropical cyclones can form and strengthen over land because they are dependent on strong contrasts in temperature across a frontal zone rather than on a large source of water vapour (as in the case of tropical cyclones). The contrast between cold, dry winter air over continental interiors and warmer, moist air over the sea (§ 5.2.3) creates favourable conditions for the formation of extratropical cyclones, and so extratropical cyclones tend to be strongest in the late-winter months when the pole is at its coldest and temperature gradients are greatest (see §§ 5.2.2.2 and 5.3.4). The severe winter snowstorms that are common at this time of year in much of the mid-latitudes are perhaps the most hazardous examples of such extratropical cyclones. However, extratropical cyclones have a less marked dependency on season than tropical cyclones, and can occur at any time of year.

As winds circulate around the cyclone, the front warps, into something like the shape shown in Figure 4.6. The heaviest rain (or snow) usually occurs to the

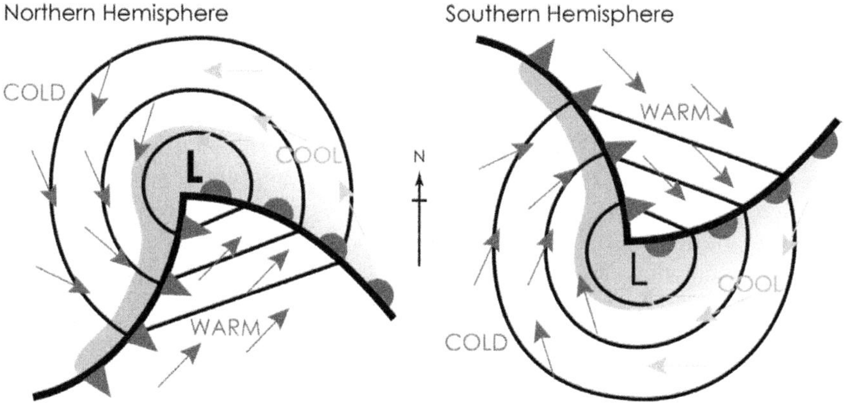

FIGURE 4.6 Typical structure of a mature extratropical cyclone. The arrows indicate the near-surface wind direction, and the near-concentric circles represent lines of equal air pressure, with 'L' indicating lowest air pressure in the centre. The shading indicates areas of rainfall and the thick lines indicate the fronts (the bobbled line is the warm front, and the pointed line is the cold front). North is marked by the crossed arrow.

southwest of the cyclone in the Northern Hemisphere (or to the northwest in the Southern Hemisphere) where the cold winds blowing from the pole meet the warmer air on the equatorward side of the cyclone (Figure 4.6). This line of transition is known as a cold front, and there is often a noticeable drop in temperature once the front has passed. Substantial rain can also occur to the east where the warm air on the equatorward side moves north to meet the cold air. This line is known as a warm front, and it usually brings less heavy but more continuous rain. When the warm front passes, there is often a noticeable increase in temperature. As with tropical cyclones, flooding can occur, or large amounts of snow can fall, if the cyclone is slow-moving.

Extratropical cyclones are less symmetric than tropical cyclones not only in their intensity of rainfall, but also in their areas of strongest winds. In the Northern Hemisphere, the strongest winds are usually to the northwest (southwest in the Southern Hemisphere) of the centre. These winds are typically weaker than in tropical cyclones because there is less energy from latent heat. However, extratropical cyclones are generally much larger because the fronts between the warm and cold air can extend for thousands of kilometres. The largest recorded tropical cyclone was Super Typhoon Tip in 1979, which was about 2220 km in diameter; for reference, most tropical cyclones are smaller than 1000 km in diameter. In contrast, 2000 km is an average size for an extratropical cyclone, and they can approach 5000 km in diameter.

Smaller-scale cyclones that last only a few days sometimes occur in very high latitudes, and mainly in winter. These polar lows form because of land-sea temperature differences (§ 5.2.3) whereas the extratropical cyclones described above are formed by a temperature difference between more tropical and more polar latitudes. The less frequent summer polar lows are important for breaking up the thinning sea ice: year-to-year variability in sea-ice extent is affected by the number and strength of these polar lows. The winter polar lows can bring exceptionally cold air, most frequently into northern Eurasia, and occasionally into Japan and Canada. In Mongolia, for example, where a high proportion of the populations have livelihoods that are almost entirely dependent on their animals, these events can be devastating, causing both economic losses and severe food security crises. Modern societies can also be severely disrupted by extreme cold events with air traffic delays, school closures and power outages. In many countries there are strong extreme cold weather–mortality associations. However, the importance of low temperatures in driving elevated seasonal winter mortality in countries where heating is largely available (e.g., USA and France) is unclear.

4.2.8.3 Tornadoes

Tornadoes are similar to cyclones, in that they have a core of low pressure around which winds blow; but they occur on a much smaller scale (usually not more than 2 km across and generally lasting less than ten minutes), and form within an existing thunderstorm. The vast majority of tornadoes occur in the USA, but they do also occur in many other parts of the globe where violent thunderstorms frequently form.

The most destructive winds are from tornadoes rather than tropical cyclones. However, comparisons are difficult because tornadoes are short-lived and can be fast-moving, so their maximum wind speed is recorded (a challenge in itself) instead of the sustained wind speed (§ 4.2.4), which is used to measure tropical cyclone strength. As examples, wind speeds in the strongest tornadoes have exceeded speeds of 480 km.h^{-1}, whereas sustained wind speeds above 300 km.h^{-1} in tropical cyclones are exceptionally rare. Hurricane Irma, which devastated the Caribbean in September 2017, had the highest recorded sustained winds of any Atlantic hurricane (> 295 km.h^{-1} over 37 hours). Super Typhoon Tip (1979) did generate sustained winds of 305 km.h^{-1}, but these were short-lived.

4.3 How can climate be summarized?

4.3.1 If weather is what we get, what should we expect?

Most people have a reasonable sense of what 30 °C (or 86 °F) feels like. However, while one person may consider 30 °C warm, another may consider it extremely hot, depending largely on where they each live and on the time of year. To interpret a given temperature we compare it to the temperatures to which we each are acclimatized. Herein lies a difference between weather and climate. A weather forecast will tell you what the temperature may be tomorrow in °C, but if you want to know whether it is going to be unusually hot or cold, you have to refer to the climate. The climate provides a reference as to what kind of weather does occur here at this time of year, as distinct from what specific weather may occur tomorrow.

By comparing the weather forecast with this reference – the climatology – we can identify not only whether it is going to be unusually hot or cold, but also how unusually hot or cold, or even whether record temperatures may be set. A climatology provides far more information than just the average weather conditions.

A suitable climatology encompasses previous years' observed weather conditions for a particular location and time of year (e.g., month, season). Since the weather in each year is unique, we can hopefully obtain a reasonable indication of typical weather conditions and the range of possibilities with data compiled over many years. Climates change and fluctuate, and so a climatological period needs to be selected that is sufficiently up-to-date so that the first few years in the period are a reasonable indication of the types of weather that still occur. However, the climatological period needs to be long enough to avoid an unrepresentative sample. For most purposes, climatologists have settled on using a recent 30-year period as a suitable reference and this is the standard in most countries. In a few countries, there are insufficient data (see Case Study 6.1), and some compromises have to be made. For many engineering purposes and for risk mapping, or other analyses of extremes, a period longer than 30 years may be required, but there may be limitations imposed by data availability and quality, and by representativeness in the context of climate change. Similarly, because temperatures have been increasing in many parts of the world (see Chapter 9) even a recent 30-year period can include years that are unrepresentative of more current and likely near-future conditions. However, there are no revised international guidelines for defining an appropriate climatological period for temperature, and so a recent 30-year period remains the standard.

Although there is general agreement on using 30 years for a climatology, there are some inconsistencies in which 30 years are chosen. If the most recent 30 years were used, the climatology would change a little every year. Such changes can be confusing and may not be viable if there are delays in collecting all the data for the last year anyway. So climatologies are commonly updated every ten years. The current climatological period is 1981–2010, which will be updated to 1991–2020 in 2021. For monitoring of El Niño and La Niña (for which sea temperatures rather than air temperatures are used; see Box 5.1), the climatological period is updated every five years, and so the current climatological period is 1986–2015. In a few cases, including at some of the Regional Climate Outlook Forums (see § 8.3.2), the climatological period is updated only every 30 years. For these climatologies, the current period is 1961–1990, which should be updated to 1991–2020 in 2021 or soon after. Given strong positive temperature trends, current seasonal temperature forecasts almost always indicate warming compared to 1961–1990. This lack of discrimination limits the value of these forecasts to decision-makers.

4.3.2 Aggregating weather data

Measurements of the meteorological parameters discussed in § 4.2 are recorded either as instantaneous values (e.g., air pressure), accumulations over limited periods (e.g., rainfall), averages (e.g., sustained winds or mean temperatures) or extremes (e.g., maximum and minimum temperatures) over 24-hour periods or less. So how can these measurements be represented over climatological periods or other longer periods?

For periods longer than a day, temperatures are usually averaged (although some temperature indices accumulate daily values; see Box 4.2). Rainfall is measured as an accumulation over 24 hours, and is often accumulated for longer periods rather than presented as an average daily accumulation. For example, the average rainfall for New York in July is about 100 mm per month, or a little more than 3 mm per day. Units of average millimetres per day may give the misleading impression that it rains about 3 mm every day in July, whereas typically fewer than half the days receive any rain at all so that it rains about an average of 8 mm per day in July when rain does occur. Rainfall can be accumulated for periods longer than a month, such as three-month seasons or even a year, but as soon as discontinuous periods, or periods longer than a year, are considered (e.g., 30 years' worth of July totals) the accumulations are averaged. For example, the climatological June–August average rainfall is the average of the 30 individual 92 daily accumulations.

Climatological averages are used extensively, but are often misleading. For example, about half the time it is warmer than the average temperature, and the other half it is colder, but similar conclusions can be quite inaccurate for rainfall or for wind speeds. The difference depends on the statistical distribution of the values in the climatological data. Monthly and seasonally averaged temperatures are normally-distributed for most places and seasons, while daily temperatures may be slightly skewed. The skewness can be positive or negative, and is not marked either way, but may be sufficient to be important in some statistical analyses in, for example, coastal areas. For rainfall, departures from normality can be extreme. Daily values nearly everywhere are well approximated by negative exponential or mixed negative exponential distributions (left part of Figure 4.7), and for most modelling work rainfall occurrence is represented separately from rainfall amount because of high frequencies of dry days. The distribution of monthly, seasonal and annual rainfall accumulations depends partly on location and time of year, but some degree of positive skewness is likely to be evident in most cases (right part of Figure 4.7). The skewness is strongest in dry locations and times of the year.

The prevailing and predominant winds (§ 4.2.4) are widely used to summarize directions. For climate analyses, atmospheric scientists divide (the horizontal part of) winds into two components: the north-south ('meridional') component and the east–west ('zonal') component.

4.3.3 How hot is hot? When does dry mean drought?

Climatologists frequently use the word 'anomaly', which, it must immediately be noted, means something quite different when used by a health expert. To a climatologist an 'anomaly' is not something abnormal or an indication of an error in the data; instead it is a difference from average. Temperatures are widely reported

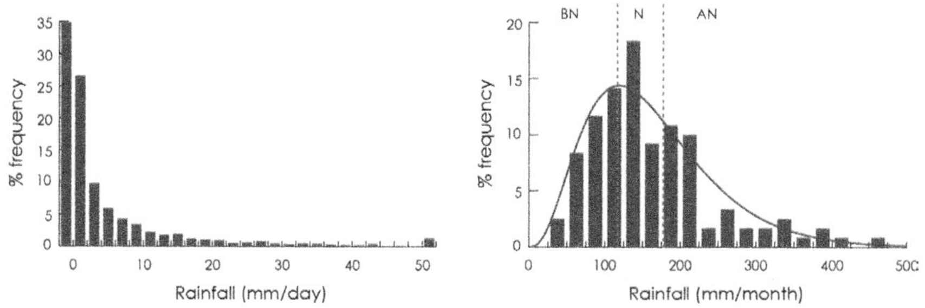

FIGURE 4.7 Frequency distributions of daily (left) and monthly (right) rainfall accumulations for Barbados for the wet season (August–November) 1981–2010. The dotted vertical lines on the right diagram indicate the lower and upper terciles and the corresponding categories, and the solid curve indicates a gamma distribution fit to the data.

as anomalies. For example, 2016 was 0.94 °C hotter than the 20th century average (100-year, rather than 30-year, climatologies are often used in the context of global average temperature monitoring). Most people have some sense of what a given temperature anomaly might feel like (although probably little sense of what that same anomaly means when averaged over the whole globe or over a large area, or even over a long period of time).

Compared to temperature, rainfall anomalies are much harder to interpret, in part because most of us have a much weaker impression of what a typical amount of rainfall is compared to what a typical temperature is. Is 10 mm (about 0.4 inches) in one day a lot of rain? What about 200 mm (about 8 inches) in three months? Unless you come from a desert or are currently in the dry season, those questions may not be easy. One simple solution is to express rainfall as a percent of, rather than difference from, average. New York received 63 mm of rain and snow in February 2017 (see Box 4.2); that was 6 mm more than average, which represents a little over 10% extra. There are on average about nine wet-days per month in February in New York, so the extra 10% represents approximately one extra day's worth of rain / snow.

Even as a percent of average, it is not immediately obvious whether February 2017 was unusually wet in New York. 'An extra 10%' may be a bit clearer than 'an extra 6 mm', but even percentages can be hard to interpret: in very dry climates, a little bit of extra rain can translate into a very large percentage. The problem is compounded by differences in variability of climate (§ 5.3.5). For example, the most severe drought on record in New York was in 1965, when rainfall was 58% of average, and the most recent drought was in 2001, when there was 80% of average rainfall. In comparison, during the drought in East Africa in 2011, large areas received less than 25% of average rainfall, and still did not break their records.

To some extent, there is a similar problem with temperatures: in February 2017, New York was 3 °C hotter than average. In the tropics, where temperatures generally do not change much from day-to-day, a month of 3 °C excess heat per day would be exceptional. Temperature anomalies cannot be converted to percentages of average (unless one wishes to work in the Kelvin scale) because of the arbitrary value for zero degrees, but regardless of whether anomalies or percentage departures are used, temperature and rainfall values are frequently difficult to interpret. Often we want to know how unusually mild the winter has been or how unusually severe the drought was, and so we need to compare the event of interest not just with the average conditions, but also with previous mild winters or with past droughts.

Perhaps the simplest way to assess how unusual a recent event has been is to compare it with previous record values. For example, globally, 2016 was the warmest year on record. The same was true of 2015 and 2014, so the global temperature record has been broken three years in a row. Records can be used in exactly the same way for rainfall and other variables. Records are generally calculated using all available data, although, for less extreme cases, comparisons may be made with the most recent years: for example, the 2016/2017 winter was 'the driest in 20 years' in much of the United Kingdom. Alternatively, if a record is not quite broken, the ranking can be indicated: for example, 2016 was the tenth warmest year on record in France.

The ranking example for France's 2016 country-averaged temperature is similar to how climatologists define categories. Categories are usually defined using the climatological (30-year) data, and the most common practice is to define three categories ('above-normal', 'normal' and 'below-normal') so that there are equal numbers of years in each one (see Box 8.3). Typically, the range of the 'normal' category is narrow (see example in Figure 4.7, where the 'normal' category is bounded by the two vertical dotted lines), and so 'above-normal' and 'below-normal' may not be particularly extreme.

Standardized anomalies (the anomaly divided by the standard deviation) can be meaningful if the data are normally distributed, but for rainfall, such an assumption is often invalid, and results can become misleading. As a solution, rainfall data are sometimes transformed to an approximately normal distribution by using a gamma or Pearson Type-III distribution fit. The resultant Standardized Precipitation Index (SPI)[i] is widely used for drought monitoring. As a drought monitoring index the SPI is calculated using anywhere between the most recent 1 and 24 months' rainfall accumulation. The SPI values are standard normal deviates, and so return periods (how frequently a drought of a given intensity is expected to occur) can be calculated using standard normal distribution tables. Fixed SPI thresholds are widely used to indicate varying levels of drought severity, although the thresholds may vary from country to country (Table 4.1).

Unfortunately, the SPI is not always implemented appropriately for every application. For regions without a strong seasonal cycle in rainfall, drought can be monitored effectively by measuring the SPI on rainfall for the past few months (typically 3 or 6), but this practice is often implemented indiscriminately, including in areas with strong seasonal cycles. In areas where most of the annual rainfall is received

TABLE 4.1 Standardized Precipitation Index (SPI) thresholds and corresponding return periods (in years) for droughts of varying severity, as defined by the World Meteorological Organization (WMO) and the United States Drought Monitor (USDM)

Drought severity	*WMO*		*USDM*	
	SPI value	*Return period*	*SPI value*	*Return period*
Mild / Abnormally Dry	0.0 to −1.0	1 in 2	−0.5 to −0.7	1 in 3
Moderate	−1.0 to −1.5	1 in 10	−0.8 to −1.2	1 in 5
Severe	−1.5 to −2.0	1 in 20	−1.3 to −1.5	1 in 10
Extreme	< −2.0	1 in 50	−1.6 to −1.9	1 in 20
Exceptional			< −2.0	1 in 50

during a wet season of a few months, measuring a three-month SPI over the dry season would not give a particularly informative indication of whether drought conditions are occurring – an abnormally dry dry-season may have only minor impacts.

This problem can be avoided by using 12- or 24-month SPIs, for example, although in regions with more than one rainy season or with no marked rainy season a 12-month index may be too long. Therefore, when measuring or forecasting drought, care must be taken to select an appropriate number of weeks or months to account for the local timing and duration of rainy seasons. The Weighted Anomaly Standardized Precipitation (WASP) Index[18] is an attempt to simplify such questions; it weights each month's rainfall anomaly based on how wet or dry that month is typically. A dry-season month will be given low weight, and if all the months selected are in the dry season the index will not give a strong value unless it was exceptionally dry. However, the WASP does not convert as easily to return periods as does the SPI.

4.4 Conclusions

Some of the important principles for understanding climate are: 1) the air is heated primarily from the bottom by Earth's surface, not from the top by the sun; 2) air cools as it rises; 3) warm air can hold a lot more moisture than cold air; and 4) water does not change temperature very easily. These principles help to explain how rain occurs, why there is often a notable change in temperature after rainfall in the extratropics, where tropical cyclones form, etc. These principles are referred to frequently in the subsequent chapters. In the following chapter they are used to help explain why climate varies in space and time.

Note

i www.wamis.org/agm/pubs/SPI/WMO_1090_EN.pdf.

References

1 Lamb, H. H. *Climate, History and the Modern World*. 2nd edn (Routledge, London, 1982).
2 Larson, L. W. & Peck, E. L. Accuracy of precipitation measurements for hydrologic modeling. *Water Resources Research* **10**, 857–863 (1974).
3 Ghipponi, P., Darrigol, J., Skalova, R. & Cvjetanović, B. Study of bacterial air pollution in an arid region of Africa affected by cerebrospinal meningitis. *Bulletin of the World Health Organisation* **45**, 95–101 (1971).
4 Shaman, J., Pitzer, V. E., Viboud, C., Grenfell, B. T. & Lipsitch, M. Absolute humidity and the seasonal onset of influenza in the continental United States. *PLOS Biology* **8**, e1000316, doi:10.1371/journal.pbio.1000316 (2010).
5 Service, M. W. Effects of wind on the behaviour and distribution of mosquitoes and blackflies. *International Journal of Biometeorology* **24**, 347–353 (1980).
6 Benton, B., Bump, J., Seketeli, A. & Liese, B. Partnership and promise: evolution of the African river-blindness campaigns. *Annals of Tropical Medicine and Parasitology* **96**, 5–14 (2002).
7 Boatin, B. The onchocerciasis control programme in West Africa (OCP). *Annals of Tropical Medicine & Parasitology* **102**, 13–17 (2013).
8 Boakye, D. A., Back, C., Fiasorgbor, G. K., Sib, A. P. P. & Coulibaly, Y. Sibling species distributions of the Simulium damnosum complex in the West African Onchocerciasis control Programme area during the decade 1984–93, following intensive larviciding since 1974. *Medical and Veterinary Entomology* **12**, 345–358 (1998).
9 Baker, R. H. A. *et al*. Progress in controlling the reinvasion of windborne vectors into the western area of the onchocerciasis control program in West-Africa. *Philosophical Transactions of the Royal Society B: Biological Sciences* **328**, 731–750 (1990).
10 Thomson, M. C. *et al*. The unusual occurrence of savanna members of the Simulium damnosum species complex (Diptera: Simuliidae) in southern Sierra Leone in 1988. *Bulletin of Entomological Research* **86**, 271–280 (1996).
11 Walsh, J. F., Molyneux, D. H. & Birley, M. H. Deforestation – effects on vector-borne disease. *Parasitology* **106**, S55–S75 (1993).
12 Wilson, M. D. *et al*. Deforestation and the spatio-temporal distribution of savannah and forest members of the Simulium damnosum complex in southern Ghana and south-western Togo. *Transactions of the Royal Society of Tropical Medicine and Hygiene* **96**, 632–639 (2002).
13 Hougard, J. M. *et al*. Criteria for the selection of larvicides by the onchocerciasis-control-program in West-Africa. *Annals of Tropical Medicine and Parasitology* **87**, 435–442 (1993).
14 Dee, D. P. *et al*. The ERA-Interim reanalysis: configuration and performance of the data assimilation system. *Quarterly Journal of the Royal Meteorological Society* **137**, 553–597 (2011).
15 Heblich, S., Trew, A. & Zylberberg, Y. *East Side Story: Historical Pollution and Neighborhood Segregation* (Mimeo, New York, 2016).
16 Solomon, S. Progress towards a quantitative understanding of Antarctic ozone depletion. *Nature* **347**, 347–354 (1990).
17 Landrigan, P. *et al*. The Lancet Commission on pollution and health. *Lancet* **391**, 462–512, doi:10.1016/S0140-6736(17)32345-0 (2017).
18 Lyon, B. & Barnston, A. G. ENSO and the spatial extent of interannual precipitation extremes in tropical land areas. *Journal of Climate* **18**, 5095–5109 (2005).

5

CLIMATE VARIABILITY AND TRENDS

Drivers

Simon J. Mason
Contributors: Ángel G. Muñoz, Bradfield Lyon and Madeleine C. Thomson

Free lords, cold snow melts with the sun's hot beams
Henry VI, Part II *by William Shakespeare*

5.1 Introduction

Chapter 4 describes the basic components of weather and climate, and a common theme throughout is that there is considerable variability in space and in time. In this chapter, we start by describing and explaining how climate varies by location, by considering the effects of altitude, latitude and other aspects of geography on the climate. We then examine how climate varies over time, starting with differences between night and day, describing the seasons and how they are affected by location, and then describing how and why climate varies from year-to-year and at even longer timescales. Based on what we have learned in Chapters 1–4 the connection of climate to the spatial and temporal risk of infectious diseases, malnutrition or hydro-meteorological disasters can now be made.

5.2 How does climate vary spatially?

The average temperatures and rainfall for January and July across the globe are shown in Figure 5.1 and Figure 5.2. Why are some places hotter, or drier, than others? The most important factors affecting the spatial variation of climate are:

- Altitude
- Latitude
- Contrasts between the effects of land and sea
- Different land surface conditions

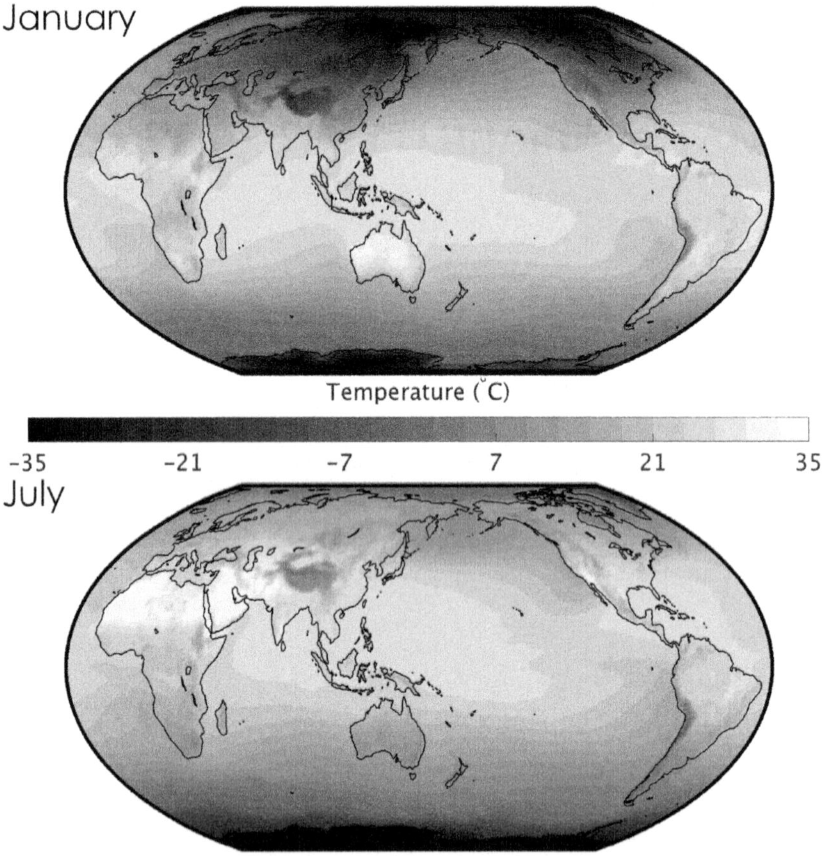

FIGURE 5.1 Average temperatures for January (top) and July (bottom) 1981–2010. *Data source*: ECMWF Interim Reanalysis, for 1981–2010[1]

5.2.1 Climate and Altitude

5.2.1.1 Temperature and altitude

An important lesson of Chapter 4 has been that the air is heated from Earth's surface rather than directly by the sun. This distinction has profound effects on how temperatures vary horizontally and vertically. Consider first, vertical changes in temperature. Even from a cursory inspection of Figure 5.1, an effect of altitude on temperatures is evident from the relatively cold temperatures over the Himalayas, the Andes and other high mountain ranges. It is easier to see the effects of altitude by looking at how temperatures change above a specific location (Figure 5.3; in this case, Brookhaven, NY, but the general pattern is similar in most locations). There are two main features:

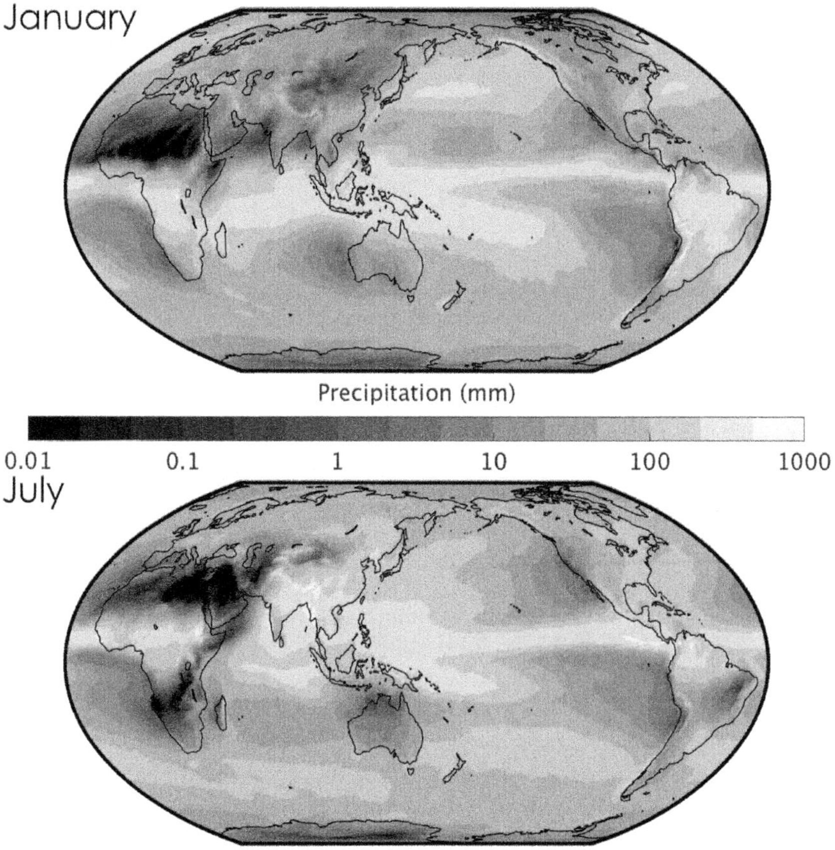

FIGURE 5.2 Average rainfall for January (top) and July (bottom) 1981–2010.
Data source: ECMWF Interim Reanalysis, for 1981–2010[1]

- Temperature decreases rapidly at a near-constant rate for about the first 10 km (jet airplanes fly at about this altitude, where you may have noticed recordings of outside temperatures of around –60 °C or –70 °C).
- Above about 10 km, temperatures decrease more slowly, and start to increase above about 20 km.

The air is warmest near the surface because that is where the air is heated by the Earth, but the secondary peak in temperatures at about 25 km altitude is where the sun can heat the air up directly because of the presence of ozone at these altitudes. Ozone is one of the few gases that directly absorbs the sun's rays – specifically ultraviolet radiation (§ 4.2.6) and so at these high altitudes the sun heats the ozone directly, which, in turn, heats up the gases in the immediate vicinity.

The rapid temperature decrease in the first 10 km above the surface is a direct effect of a decrease in air pressure at higher altitudes (§ 4.2.7). Air pressure decreases

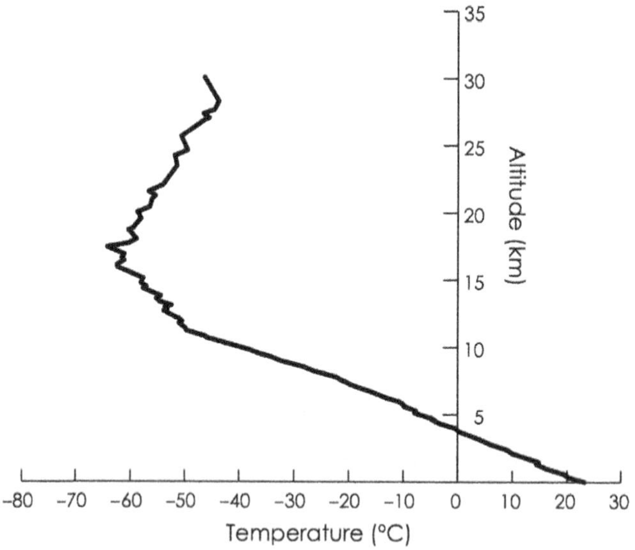

FIGURE 5.3 Temperature as a function of altitude over Brookhaven, NY, at 08h00 local time on 3 August 2017
(*Data source*: https://ruc.noaa.gov/raobs/)

with elevation because there is less weight of air pressing down from above. As incredible as it may sound, if you stand on the beach at the sea there is about a tonne of air pressing down on you. At sea-level the air is squashed because of all that weight, but higher up there is less weight of air above and so the air is less dense and can expand into more space. For an equivalent reason, we do not put our tomatoes in the bottom of the shopping bag, otherwise the heavier items on top will squash them. The amount of air decreases by about 11% for every kilometre in altitude.

Because of this decreasing air pressure, air expands as it rises, while as it descends it compresses; but, by expanding, the air cools, and by compressing it warms up. This cooling and heating effect can be reproduced with a simple experiment. If you blow on your hand, your breath feels cold because it is forced through a narrow hole in your mouth and then expands as it gets into the open air; but huff and your breath feels warm because it is not expanding into the open air. Similarly, if you pump up a bicycle tyre you may notice the pump gets hot because of the air compression, but the air escaping from the valve or through a puncture feels cold.

The cooling of air as it expands when it rises occurs at a predictable rate known as the *lapse rate*. As a reasonable rule of thumb, the temperature decreases about 1 °C every 100 m, which is approximately the rate of cooling you would feel when ascending in a hot air balloon. Conversely, air warms by about 1 °C every 100 m as it sinks. That lapse rate is the same in hot and cold air, and so it is the same in the tropics and the extratropics. However, there is an important complication: it is not the same in humid and dry air. Air only cools at this rate when it is unsaturated (i.e., the relative humidity is less than 100%).

The heat that was originally used to evaporate water is released if the air is cooled enough to condense the vapour back to water (see the discussion on latent heat in § 4.2.1). This latent heat partly reheats the air so that the air cools more slowly as it ascends once it becomes saturated. On average, the decrease is about 0.6 °C every 100 m, but the actual rate depends heavily on how much water vapour does condense, which in turn depends on the temperature. The amount of water vapour required to saturate air increases exponentially with temperature, and so hot humid air contains a lot more latent heat than cool humid air (§ 4.2.3). Therefore, as hot humid air cools, it condenses large amounts of water vapour and releases large amounts of latent heat, but cold humid air releases only small amounts of latent heat. As a result, the lapse rate is much lower in hot humid air than in cool humid air. One important effect of this lower lapse rate is that rising hot humid air is likely to stay hotter than its surroundings because it cools only a small amount with height, and therefore it will remain buoyant. The buoyancy of hot humid air because of the condensation of large amounts of water vapour is critical in forming heavy rainfall and violent storms in the tropics. It is one reason why hurricanes and typhoons, for example, form only in the tropics (§ 4.2.8).

The lapse rate indicates how quickly air cools as it rises, but that does not mean that temperatures decrease by that amount if you climb up a hill. If a wind blows up a hill, the air will cool as it ascends, but because the air is still near the surface, that cooling may be offset by heating from the hill itself. The actual decrease in air temperature uphill is affected by many factors including the orientation of the slope and exposure, the weather and the time of day. It may even be warmer up the hill if the air is clearer there or if the slope faces the sun more directly. It is important to be aware of such micro-climatic variations when temperature thresholds are being used to guide decision-making processes. Nevertheless, temperatures do generally decrease uphill. In the Himalayas and the Alps, for example, temperatures decrease uphill at a rate of about 0.4 °C to 0.7 °C every 100 m. An average rate of about 0.6 °C (about the same as the average lapse rate in moist air) is a reasonable rule of thumb. Understanding just how temperatures cool at higher altitudes is important for malaria planning (see Case Study 5.1).

The rapid decrease in temperatures in the first 10 km that is shown in Figure 5.3 is typical of most places and times of year. However, under certain weather conditions temperatures can increase for the first few hundred metres. Such an increase is known as an *inversion*.

Inversions can occur when warm air blows over a colder surface. Fog may form from this process if the cooling is enough to produce condensation. This type of fog is common in mid-latitude maritime climates (see Box 4.1 for definitions of climate regions) during the winter. It also occurs in the subtropics along the west coast of many of the large desert regions, where winds may be cooled by a cold sea surface.

CASE STUDY 5.1 ELEVATION USED IN PLANNING MALARIA CONTROL PROGRAMMES

Bradfield Lyon, University of Maine, USA and Madeleine C. Thomson, IRI, Columbia University, New York, USA

The East African highlands have long been considered desirable for human habitation because of their rich soils and low levels of infectious diseases such as malaria. Since air temperature decreases with elevation (§ 5.2.1.1), the degree days required for malaria parasite development become increasingly difficult to achieve the higher up the mountain one goes until a minimum temperature threshold is reached where transmission is no longer possible. Much of the Ethiopian highlands are above the altitude at which this threshold is reached, with low temperatures forming a natural barrier to malaria transmission. As such, elevation is used in malaria control planning by the Ethiopian Ministry of Health[2] as high altitudes (> 2000 m) are considered malaria-free. Below 2000 m and outside desert areas, the country is considered endemic for malaria and therefore included in routine control efforts. However, the association of temperature and altitude is not uniform across the country: there is a wide range of temperatures for stations at similar altitudes from different parts of the country (see Figure 5.4), while the overall relationship is close to the

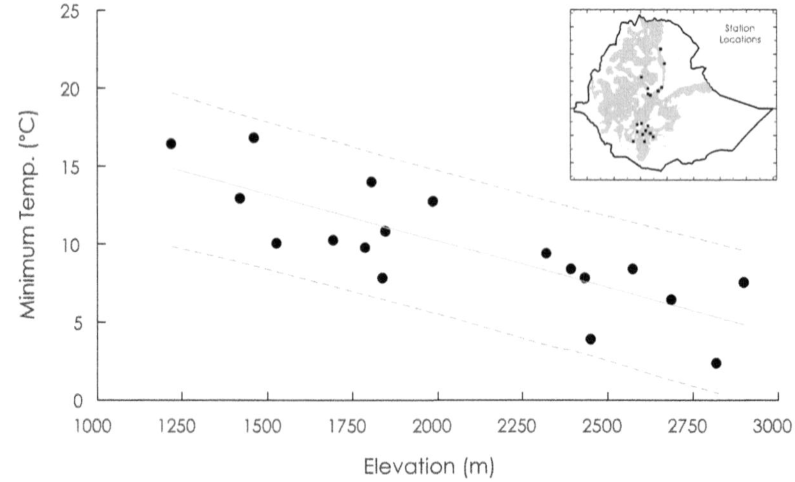

FIGURE 5.4 Climatological (1981–2010) monthly average minimum temperature (°C) as a function of elevation for 18 stations in the Ethiopian Highlands (locations shown in insert). Points indicate the lowest climatological monthly minimum temperature during the calendar year, with the solid line showing a least squares linear fit and dashed lines indicating the 95% confidence limits for predicted values

expected lapse rate (§ 5.2.1.1). A new malaria transmission boundary map was created using high resolution gridded temperature data from the National Meteorological Agency.[3] This boundary map revealed that there can be a 1000 m difference in elevation for the same minimum temperature threshold depending on where it was measured (e.g., which side of the mountain).

Some of the strongest inversions occur in the extratropics on cloudless winter nights. Such weather conditions are typical of periods of high air pressure (§ 4.2.7) where the air is descending and therefore warming. As it approaches the ground, this descending air may become hotter than the Earth's surface, especially on a clear winter night when the ground can cool down quickly. In such weather conditions, air pollution may become severe, partly because more fuel is burned for heating, but also because the inversion prevents the pollution from dispersing easily. Because the air aloft is warmer, the pollution cannot easily rise to higher altitudes where it could be dispersed by stronger winds. This cause of severe air pollution episodes is often a problem in cities that have a winter dry-season and weak winds, such as New Delhi, Santiago, Mexico City and Ulaanbaatar. Such weather conditions are also responsible for some of the most severe episodes of pollution in cities with more variable weather, such as London's Great Smog of 1952, and New York City's 1966 smog. In 2017 New Delhi had the unenviable title of the most polluted city on the planet. On 8 November 2017, the city's air quality index measured in the range of 700–1000; the US Environmental Protection Agency considers anything over 300 to be hazardous. The extreme smog resulted from smoke caused by burning of stubble on nearby farms coinciding with a temperature inversion that kept the smoke and other polluted air in the city for several days.

5.2.1.2 Humidity and altitude

Just as temperatures decrease with altitude, so does humidity. How humidity decreases depends, in part, on how it is measured (see Box 4.3), but it is sufficient to note that the amount of water vapour in the air decreases with altitude largely because of the effects of decreased temperature and air pressure. The cooler temperatures and lower humidities mean that heat stress (see Box 4.2) is unlikely to be a major health problem at high altitudes, both now and in the future.

5.2.1.3 Wind and altitude

Although heat stress at high altitudes is unlikely to be a problem, wind chill is a bigger risk, not only because of the colder temperatures, but also because wind speeds tend to increase with altitude. Wind speeds increase with altitude partly because there is less friction with the surface, and partly because air pressure gradients generally strengthen. Wind chill combined with extremely low temperatures can be exceptionally hazardous for mountaineers.[4]

5.2.1.4 Rainfall and altitude

The complicated relationship between humidity and altitude is reflected in that between rainfall and altitude. Nevertheless, some general patterns can be identified. Up to a point (discussed further below), rainfall increases at higher altitudes on the windward (and often wetter) side of mountains, and so knowledge of the prevailing wind direction (§ 4.2.4) is important for understanding mountain climates. The increase in rainfall with altitude is a result of the orographic effect (§ 4.2.2): winds blowing towards a mountain are forced to rise, resulting in cooled air, from which water vapour may condense to form clouds and possibly rainfall. On the leeward side, the winds descend the mountain, the air warms, clouds are evaporated, and so rainfall is inhibited.

The contrast between the relatively wet windward and dry leeward sides of mountains is evident across the globe. However, the rate of change in rainfall with altitude on the windward side is complicated. On the one hand, the higher up the mountain, the cooler the air becomes, and so the greater the likelihood that water vapour will condense. On the other hand, much of the water vapour at high altitudes may already have condensed and so the tops of mountains may be above the clouds. Even if there is still moisture left, the low temperatures at these high altitudes mean that only small amounts of condensation will occur because of the exponential relationship between temperature and saturation that was discussed in §§ 4.2.3 and 5.2.1.2. Therefore, rainfall starts to decrease with altitude beyond an 'elevation of maximum precipitation'. For most of the globe this elevation is somewhere between about 1 and 2 km, but in polar latitudes, the air is too cold to hold much moisture. Here rainfall (or, more typically, snowfall) is at a maximum at the base of a mountain, and decreases with elevation. In the humid tropics, the elevation of maximum precipitation tends to be on the lower side of the 1–2 km range because of the large volumes of condensation that occur at low altitudes where the air is most humid.

If nothing else is clear beyond the fact that this relationship between rainfall and altitude is complicated, you have understood the main conclusion! What generalizations, if any, can be drawn? The altitude of maximum rainfall tends to be higher in the mid-latitudes than in the tropics and high latitudes, and for similarly complicated reasons, the altitude of maximum rainfall tends to be relatively high in dry areas, in the dry season, and in the warm season.

5.2.2 Climate and latitude

5.2.2.1 Rainfall and latitude

As just noted, not much snow falls near the Poles because the air is too cold to hold much moisture. For this reason, Antarctica (the left side of Figure 5.5) is the driest continent on Earth. Antarctica is drier than the Arctic (the right side) because the South Pole is so much colder than the North (Figure 5.6). However, rainfall does

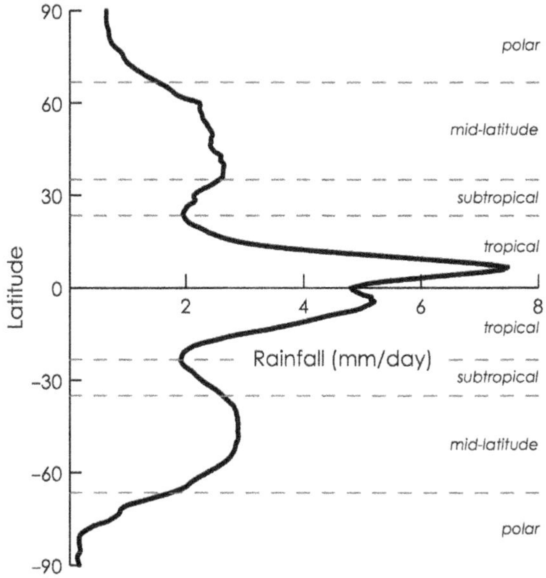

FIGURE 5.5 Rainfall (and snow) as a function of latitude (the South Pole is at the top; the North Pole at the bottom).
Data source: ECMWF Interim Reanalysis, for 1981–2010[1]

not decrease simply from the warm tropics to the dry polar latitudes (Figure 5.5): there is a band of deserts through the subtropics as well (including the Sahara in the Northern Hemisphere, and the Namib and the deserts of Australia in the Southern), which are visible in Figure 5.2. This arid belt at about 30° latitude occurs in both hemispheres, and is caused by poleward moving air from the equator sinking in the subtropics as a result of Earth's rotational effects.

Most rainfall occurs near the equator where solar heating, and therefore evaporation of surface water, are at a maximum, and where high temperature and humidity can encourage the formation of heavy rainstorms (§ 4.2.8). There is a slight local rainfall minimum on the equator itself because of a surprisingly cold sea surface in the eastern Pacific Ocean (Figure 5.1) which is relatively dry as a result (Figure 5.2). This pattern of a cold and dry eastern equatorial Pacific is occasionally disrupted as part of El Niño (Box 5.1).

5.2.2.2 Temperature and latitude

The South Pole is colder than the North Pole (Figure 5.6) primarily because of differences in altitude (§ 5.2.4). The North Pole is near sea-level whereas much of Antarctica is above 2 km; Antarctica's altitude contributes to at least a 15 °C cooling. The temperature difference is exacerbated by land–sea contrasts: the South Pole is straddled by the continent of Antarctica while the North Pole is straddled by the

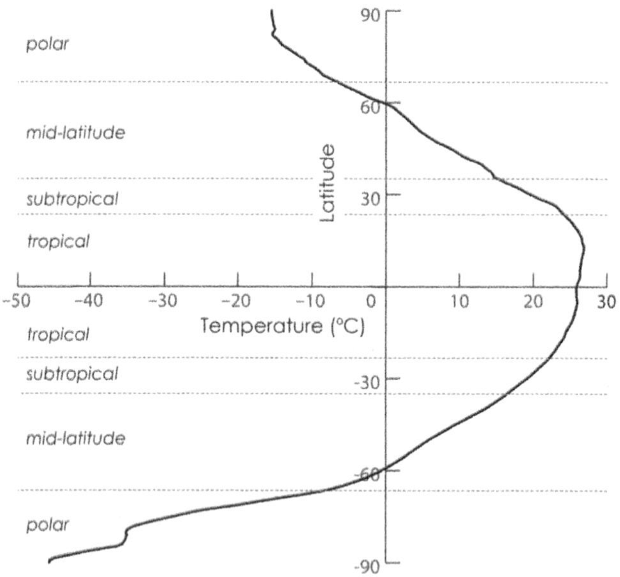

FIGURE 5.6 Temperature as a function of latitude (the South Pole is at the top; the North Pole at the bottom).
Data source: ECMWF Interim Reanalysis, for 1981–2010[1]

Arctic Ocean (see § 5.2.6 on the contrasting effects of land and sea). Areas near the equator have the highest temperatures on average through the year, and temperatures are consistently high through the year. However, the most extreme temperatures (well above 50 °C) occur away from the sea in the subtropics (specifically, in Kuwait, Libya and California), where land-surface temperatures exceeding 80 °C have been recorded. These locations have a large annual range in temperature: they experience relatively cold winter and night-time temperatures compared to at the equator (§§ 5.3.1 and 5.3.4).

Why do air temperatures decrease towards the Poles? The answer has to do with the sun's heating of Earth's surface. The air is heated by Earth's surface, but the surface temperature depends largely on the intensity and duration of the sun's heating. During the winter, because the sun does not appear at the Poles at all, it does not heat the surface for weeks or months (see § 5.3.3). During the summer, although the sun is up all day, it remains low in the sky, which has two effects:

- The intensity of the sun's heating is weak. The sun's heating is most intense when the sun is high in the sky – directly overhead. If you shine a torch directly on the ground it will make a small bright circle, but shine it on the ground at an angle and it will make a larger and dimmer oval. The torch is emitting the same amount of light, but that light spreads over a larger area simply because of the angle. The same is true of sunlight: at about 60° latitude, sunlight is spread over about twice the area it as at the equator because Earth is spherical. Near the Poles (or at any latitude in the morning, or in the mid-latitude winter), the sun is low in the sky and so it cannot heat the surface very intensely, but near the equator (or at high noon in summer) the sun is more directly overhead and its heating is more intense. Averaged over a year, the Poles receive the least amount of the sun's energy, but for a short time in the summer they do receive the maximum amount per day. Despite the low height of the sun, the Poles get more sunlight per day in their peak summer than the equator gets at any time of year. The 24-hour summer days at the Poles more than offset the low height of the sun in the sky.

- Much of the sunlight at the Poles is reflected and so does not heat the ground at all. Despite the 24-hour summer daylight in the summer much of that sunlight is reflected, and so it does not heat the surface at all (see the discussion on albedo in § 4.2.7). Therefore, the Poles remain cold year-round. The highest recorded temperature at the South Pole, for example, is well below freezing point – about –12 °C.

5.2.3 The effects of land and sea

It takes an enormous amount of energy to heat up water: about four times more than for the air, and even more than for the ground. On a sunny summer's day, you can burn your feet on the paving stones around a swimming pool, but the water stays a pleasant temperature. Occasionally there are news reports of people frying eggs on a car bonnet in summer, but never about boiling an egg in a bucket of water left in the sun. It is why water is used as a coolant in car radiators, for example – water will cool down most things (metals, air, etc.) very effectively, while heating up only a small amount itself in the process.

5.2.3.1 Effects on temperature and the seasons

Because it takes so much energy to heat water, the oceans and large lakes have a major dampening effect on climate, and even rivers and small lakes may have a detectable effect. The difference in temperature between the warmest and the coldest times of the year (Figure 5.7) is one measure of the dampening effect – where the difference is small the seasons are less extreme. This annual range in temperature exceeds 20 °C over large parts of North America and Eurasia, whereas over much of the oceans the annual range is less than 10 °C. The contrast in the annual range between land and sea is strongest in high latitudes where differences in the amount of sunlight between summer and winter are extreme because of large changes in the length of the day (§ 5.3.4). However, there is also an apparent longitudinal effect: in Eurasia and North America the western sides of the continents have much smaller ranges (i.e., milder seasons) than do the eastern sides. This east–west contrast is a result of the prevailing wind direction (§ 4.2.4), which is from the west at these latitudes. Westerly winds bring air from over the sea to the western sides of the continents, but from over the

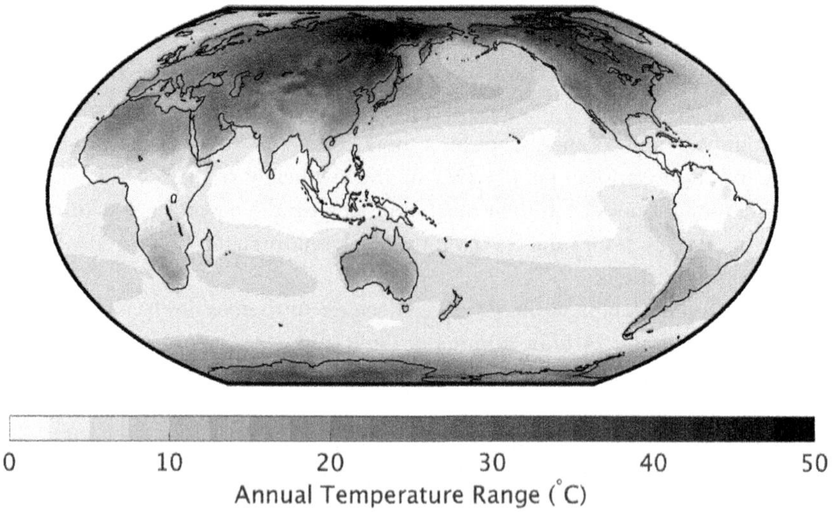

FIGURE 5.7 The annual range in temperature, calculated as the difference between the warmest and coldest mean monthly temperature. (Scale is from 0 [black] to 40 [white].)
Data source: ECMWF Interim Reanalysis, for 1981–2010[1]

land to the eastern parts. Therefore, the apparent longitudinal effect is actually an example of the effects of the distribution of land and sea.

As well as dampening the seasons, the oceans also delay them. Water heats up so slowly, that the warmest time of year at coastal areas is delayed until well after the time of strongest solar heating. Similarly, the sea cools down very slowly – once water is heated up, it stays hot for a long time – and so the coldest months in coastal areas occur well after the shortest day of the year. This delaying effect on the seasons is most evident in the mid-latitudes, where sea temperatures do not reach a maximum until about two months after the summer solstice.

5.2.3.2 Effects on humidity and rainfall

Because the sea and major lakes are the primary sources of moisture in the atmosphere, humidity is generally higher in coastal areas than it is inland. However, the influence of the oceans on the distribution of humidity and rainfall is much more a reflection of sea-surface temperatures than it is of distance to the coast. Where the sea is cold, rainfall and humidity are likely to be low, but near warm oceans humidity is likely to be high and abundant rainfall may occur. This effect of sea-surface temperatures on rainfall is most evident in the subtropics (at around 30° latitude; see Figure 5.1): here the oceans are much colder on their eastern side than on their western side because of how ocean currents are affected by temperatures and Earth's rotation. The amount of water vapour that air can hold increases exponentially with temperature, which means that over warm oceans the air can hold much more moisture than over cold oceans. As a result, the western side of continents (i.e., bordering the colder eastern ocean) tend to be much drier than the eastern side (bordering the warmer western side of the ocean).

Land–sea contrasts are important in the formation of the monsoons (see Figure 4.3). In the summertime, the land heats up more quickly than the sea and so the hotter air over the land becomes less dense. Winds will blow inland from the cooler sea to compensate for the difference in density. In winter, the land cools down more than the sea, and so the colder air over the land becomes denser and will blow towards the sea. There is therefore a strong seasonal contrast in the prevailing winds, and typically in the rainfall too: the winds from the sea are likely to bring humid rainy weather, but those from the land bring dry conditions. The most pronounced example of this contrast is the Indian Monsoon.

Similar mechanisms can apply on a smaller-scale with major lakes to affect local climate. In temperate zones, lakes moderate the temperatures of the surrounding land, cooling the summers and warming the winters, such as around the Great Lakes of North America. Lakes also act like giant humidifiers, increasing the moisture content of the air. In the winter, this moisture contributes to heavy snowfall, known as 'lake effect' snow. Large lakes can also develop their own storm systems. For instance, Lake Victoria is prone to deadly night-time storms that kill thousands of fishermen annually. Forecasting lake storms hours or days in advance has the potential to save many of these lives.

5.2.4 The effects of land-surface type

5.2.4.1 Urban heat islands

Although the contrast between land and sea is the most important effect of Earth's surface on climate around the globe, other contrasts are also important. The importance of snow and ice formation is discussed in §§ 4.2.7 and 5.2.2. Another contrast that has important local effects is that between urban and rural areas. In large urban areas (with populations of about one million or more), the construction materials, the shapes of buildings and the impacts of human behaviour can increase the temperature by around 1 to 3 °C on average. The areas impacted by these increases in temperature are called urban heat islands. The heat island usually is strongest towards the centre of the city, but local land use patterns modulate the heat island significantly, and so temperatures can differ by a few degrees within only a few blocks. The heating effect is strongest at night and in the cold season (and is strengthened further if any snow is removed manually) when warming of as much as 12 °C has been measured, and significantly contributes to heat wave risk during the hot season (Case Study 7.2). The heat island combined with various air pollutants can also increase rainfall over, and downwind of, cities. (The pollutants increase rainfall because water vapour condenses more easily into large drops – heavy enough to fall as rain rather than just form clouds – around particles.) The effect on rainfall is strongest in the hot season.

5.2.4.2 Deforestation

There is frequent mention of the contribution of deforestation to climate change. Deforestation does affect global climate primarily because of the release of large amounts of carbon into the atmosphere from the trees and the soil, rather than through large-scale impacts of changes in the Earth's surface. It can also have an important effect on local hydrology – deforestation can contribute to increased flooding – but the impacts on climate are often exaggerated.

Although deforestation tends to impact local hydrology more than local climate, some local climate impacts can occur. Small-scale deforestation often results in an increase in local rainfall because of increased surface heating – in a similar way to the urban heat island effect on rainfall. Large-scale deforestation, on the other hand, can result in decreased rainfall because there is less moisture available to evaporate from the soil and vegetation.[5] A decrease in evaporation from the land may be more important than one might first guess: about 40% of rainfall that falls over the land is evaporated from the land rather than the sea, and that proportion reaches 70% in parts of the Amazon forest. A large amount of the rainfall over the land is therefore 'recycled' – it previously fell as rain and has not yet returned to the sea.

The rainfall decrease after large-scale deforestation, combined with the reduction in moisture in the soil and vegetation canopy, have important effects on surface temperature. Less energy is needed to heat and evaporate water, and so the ground can be heated up more quickly even though the removal of the dark vegetation cover means that more sunlight is reflected. The increased surface temperature raises the air temperature, and so large-scale deforestation can result in a hotter and drier local climate.

5.2.5 Climate and spatial scale: How big is a heat wave, or a drought?

In the previous sub-sections we have seen how climate varies over large distances – at different latitudes, with proximity to the coast, or on the different sides of a mountain – but how does climate vary over shorter distances? Is the next farm or town experiencing similar weather and climate conditions as are occurring here? The West African story *Anansi and the chameleon*[6] correctly indicates that it is quite possible for the next farm to remain dry even when it is raining heavily over your farm. However, the situation may not be as uncertain for sunny weather: if your farm is experiencing drought, it is quite likely that the next farm is experiencing the same drought. Drought is a climate condition (one day of sunny weather does not constitute a drought), and similar climate conditions occur over larger areas than do similar weather conditions.

5.2.5.1 Spatial scales of temperature

Although temperatures can vary by many degrees in short distances (because of terrain or because of marked land-use differences in cities, for example), temperature anomalies (how unusually hot or cold it is) tend to be similar over large areas, extending to about 1500 km almost regardless of the timescale of interest. Any small-scale differences in temperature anomalies are quickly smoothed over by changes in wind direction because air density, and hence winds, are strongly affected by temperatures (see § 4.2.4).

In the extratropics, temperatures are strongly affected by the large-scale wind direction because climatological temperature differences are large over distances of only a few hundred kilometres (see Figure 5.1, and § 5.2.5). Winds from a relatively warm location will increase the temperature, but the temperature will fall if the wind comes from a cold location. If the winds are northerly or southerly, the interpretation is almost always simple: winds from the poleward latitudes are likely to be cold, those from more tropical latitudes are likely to be warm. An example over the USA is shown in Figure 5.8, where northerly winds bring cold air to the central states but southerly winds bring warm air to the eastern states. For easterly and westerly winds, the effect is more complex: if the new wind is coming from a continental interior it is likely to be unusually hot during the summer, but cold

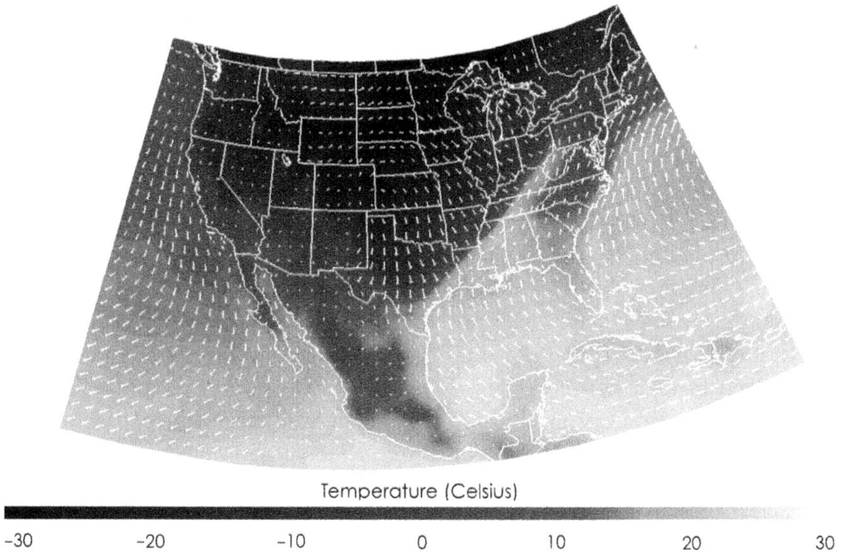

FIGURE 5.8 Example of a cold front, occurring on 26 November 2015 over part of the USA. The front lies along the boundary between the warm air to the southeast (light shading), and the cold air to the northwest (dark shading). The 2 m wind speeds and directions are shown by the arrows.
Data source: ECMWF Interim Reanalysis, for 1981–2010[1]

during the winter, and the opposite is the case if the wind is coming from the ocean. These effects of changes in the prevailing wind on mid-latitude temperatures apply day-to-day, and also year-to-year, and are inherently large-scale; in contrast, any highly localized or short-lived change in wind direction is likely to have only a negligible effect on temperature.

Warm and cold spells are likely to extend over large areas, and so if it is unusually cold in one town it is likely to be unusually cold in the next town also. However, in the extratropics at weather timescales there is often a narrow band (called a front) where the temperature changes markedly over a short distance (Figure 5.7), most commonly between autumn and spring. These fronts typically move at about 30 to 80 km.h^{-1}, and so they are not usually evident when data are averaged over more than a few days.

In the tropics, changes in wind direction do not have as strong an effect on temperature as they do in the extratropics, and so temperatures do not vary as much from day-to-day or year-to-year (§ 5.3). An exception is an occasional large-scale incursion of extratropical air in the cold season (if there is a cold season). Tropical temperature anomalies reflect changes in surface heating because of changes in cloudiness, and/or because of changes in ocean temperatures and currents, as in the case of El Niño (Box 5.1). The cloudiness changes can be highly localized, and on small mountainous islands temperature anomalies can differ considerably on opposite sides of the island. However, in most cases, the heating anomalies are large-scale, and so temperature anomalies are correlated over hundreds of kilometres.

5.2.5.2 Spatial scales of rainfall

Rainfall anomalies are less spatially coherent than those for temperature, and there is a strong dependence on timescale. The spatial coherence depends on the type of rainfall (§ 4.2.2). Convective rainfall is highly localized at weather timescales, such that individual rainfall events are hard to capture with fewer than two rain gauges per km^2.[7] Convective rainfall is common in the mid-latitudes in summer, and year-round in the tropics. One may not need to drive very far (or wait very long) for the rain to stop. However, convective showers do generally track the larger scale wind patterns and so a series of convective showers over a few hours or days is likely to deposit rainfall over a reasonably large contiguous area, much larger than the area rained upon at any moment. If it is not raining on the neighbour's farm now, it may well do so shortly, perhaps after it has stopped raining on yours, or perhaps in the next few days.

As its name suggests, large-scale rainfall is more spatially coherent than convective rain. However, even within large-scale cyclones (§ 4.2.8) rainfall intensity can vary considerably over very short distances. While it may be raining on your farm and your neighbour's farm, one of you is likely receiving quite a bit more rain than the other.

Although extratropical cyclones can be thousands of kilometres in diameter, the weather is very different in different parts of the cyclone. Cyclones are not a large-scale organization of one type of weather; instead, they are a large-scale organization of air circulation involving different types of weather. Each cyclone is unique, but as a simple generalization the following types of weather are typical. Ahead of the warm front (Figure 4.6), there is usually continuous cloud cover, and possibly light to moderate continuous rain or snow. Behind the warm front, the air feels more humid, and there may be patchy clouds. The cold front often brings heavier rain or snow with thicker clouds than in the warm front.

What can we conclude about the spatial coherence of rainfall? The amount of rainfall is highly localized at most timescales, and correlations beyond about 100 km occur only at seasonal scales. For rainfall occurrence (at the daily timescale) or frequency (for longer timescales), spatial coherency is stronger, but is usually limited to about 450 km. This is a sobering thought given that it is not uncommon in analyses of rainfall and health outcomes to use meteorological data from stations tens or even hundreds of kilometres from the site of origin for the health data.

5.3 How does climate vary temporally?

Weather and climate can vary considerably at almost all timescales, but the amount they do vary differs considerably from place-to-place, and at different times of the year. In the following sections, the effects of the time of day and the season are described and explained. Reasons why the climate is not the same every year are considered in § 5.4.

5.3.1 How does the time of day affect the climate?

5.3.1.1 Temperature

Perhaps the most immediately noticeable difference in climate conditions is the contrast in temperature between night and day. Variations through a day are called diurnal variability. The minimum temperature most frequently occurs at about sunrise (Figure 4.1), but even though the sun's heating is strongest at high noon (ignoring any complications from cloudiness), the hottest time of day may not occur until mid- or even late-afternoon because it takes time to heat up Earth's surface. The cycle of heating and cooling is not perfectly symmetrical, and so the mean temperature is generally slightly less than the average of the minimum and maximum.[8]

In humid conditions, perhaps near the coast or when the air is unusually humid, the maximum temperature occurs later than it does in more arid conditions because of the slowness of water to heat up (§ 5.2.3). In the tropics, the change in temperature during the day almost always follows a similar pattern to that shown in Figure 4.1. However, in the extratropics, the change in temperature can be strongly affected by changes in wind direction (perhaps associated with the passing of a cold front similar to that shown in Figure 5.8; § 5.2.5), and so it is possible for temperatures to decrease during the morning, or increase during the night. Nevertheless, the average diurnal cycle in the extratropics is similar to Figure 4.1. The diurnal

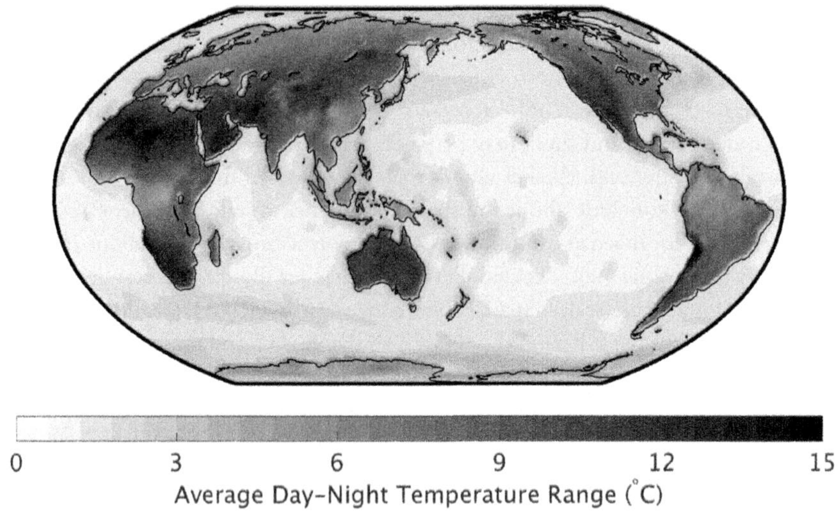

FIGURE 5.9 Average daily-range in temperature for 1981–2010.
Data source: ECMWF Interim Reanalysis, for 1981–2010[1]

range (Figure 5.9) varies considerably depending on humidity, including distance from the coast, and shows some similarities to the annual range (Figure 5.7). The effect of land–sea differences is much stronger for the diurnal range than it is for the seasonal range in temperatures because of the dominant effect of latitude on the annual range.

5.3.1.2 Rainfall

The diurnal variability in rainfall is more complicated than that for temperature, and depends on location (including altitude), time of year and rainfall intensity. Thus, it is possible to provide only some simple generalizations here. Because convective rainfall is dependent on heating and cooling of the air, it has a stronger dependence on the time of day than does large-scale rainfall. Convective rainfall peaks late-afternoon, at about the time of maximum temperature, but there is also a secondary peak a little after mid-night when cooling at night releases latent heat. Large-scale rainfall, on the other hand, does not have a strong diurnal cycle. Orographic rainfall often represents an enhancement of convective and / or large-scale rainfall so diurnal differences are similar to those just described. However, the diurnal cycle of rainfall is made more complicated in mountainous areas.

5.3.1.3 Winds

The time of day can have an important effect on local wind patterns. Wind turbulence is strongest during the daytime because of surface heating. Differences in heating between land and water bodies can create local land and sea/lake-breeze effects. These breezes are similar to the monsoons, but they occur on a daily timescale and at a smaller scale; they do not dominate the large-scale wind patterns in the way the monsoons do. Another example of a diurnal wind reversal can occur in mountainous regions. Air that is near a mountain is close to the Earth's surface – the source of heating – and so this air heats up and cools down faster than air at the same altitude further away from the mountain. During the day, the hotter air near the mountain may rise, creating a warm, uphill wind (and possibly causing rainfall), whereas at night the air near the mountain cools down quickly, resulting in a cold, downhill wind. Mountain valleys can therefore become very cold at night-time.

5.3.2 How long do weather patterns last?

Asking how long weather patterns last is like asking: how long is a piece of string? There is no way of providing a meaningful general answer to the question, and it can only be addressed in the specific: how long will *this* weather pattern last? Rainstorms, for example, move at a wide range of speeds or can stay stationary, and so can last for seconds to days or anything in between. Notwithstanding, a few generalizations are possible. For example, there is a relationship between spatial and temporal scale: convective rainfall, as well as being more localized than large-scale rainfall, is often more short-lived.

If large-scale weather conditions can have a significant impact on Earth's surface then they can have a noticeable effect on subsequent weather. For example, if snow settles then subsequent sunlight will be reflected rather than absorbed by the surface and so temperatures will probably decrease. If snow occurs over very large areas (10,000s km^2) weather patterns may be affected for the next few weeks. Similarly, after a period of dry conditions during the summer, the dry soil may be able to heat up to extremely high temperatures, drying the soil even further. The ground can then get even hotter because there is less energy used for evaporation of any remaining soil moisture. Therefore, the drying of the land surface is often an important precursor of heat waves. Dry air can inhibit the development of rain clouds thus allowing hot dry conditions to persist. However, the higher temperatures may also help to generate a sufficiently strong sea breeze to bring more moisture; so there is no simple relationship.

In the extratropics, persistent weather conditions are most frequently associated with 'blocking' episodes, so called because the usual eastward migration of weather patterns is inhibited by stationary weather systems. These blocking patterns are linked to persistent patterns in the jet streams, which are a major control on the development and movement of weather systems in the mid-latitudes (§ 4.2.8). Blocking can last for days to weeks. Because of some rather complicated effects of land and sea contrasts, blocking occurs most frequently in the Northern Hemisphere, and is most frequent in spring and over the eastern Pacific and Atlantic Oceans. The Chicago heat wave in 1995 and European 2003 and 2010 heat waves are all associated with blocking, but blocking can result in the persistence of any type of weather. Examples include the 2007 Texas floods, and cold winters in China in 2008, and in Europe 2005/2006 and 2010/2011.

5.3.3 What causes the seasons?

In Box 2.4 in Chapter 2 we learned of the importance of seasonality in driving health status, especially in rural populations in developing countries; but what causes the seasons in the first place? A very common misconception is that summer and winter are caused by changes in the distance from the Earth to the sun over the course of the year. The Earth is closest to the sun on about 3 January, i.e., in the Southern Hemisphere (*austral*) summer, and furthest from the sun on about 4 July, in the austral winter (Figure 5.10). The effect of the change in distance is small: the seasons in the Northern Hemisphere are slightly milder than they would be if the distance were the same all year, while those in the Southern Hemisphere are slightly more extreme. Instead, it is the dates of the solstices that are important in determining the seasons. The solstices are the dates on which either the most sunlight or the least sunlight is experienced; it is changes in the amount of sunshine during the year that cause the seasons.

The amount of sunshine received is much more strongly affected by the length of the day and how high in the sky the sun gets at high noon (§ 5.3.1). Winter is cold because the day is short and because the sun stays near the horizon. The

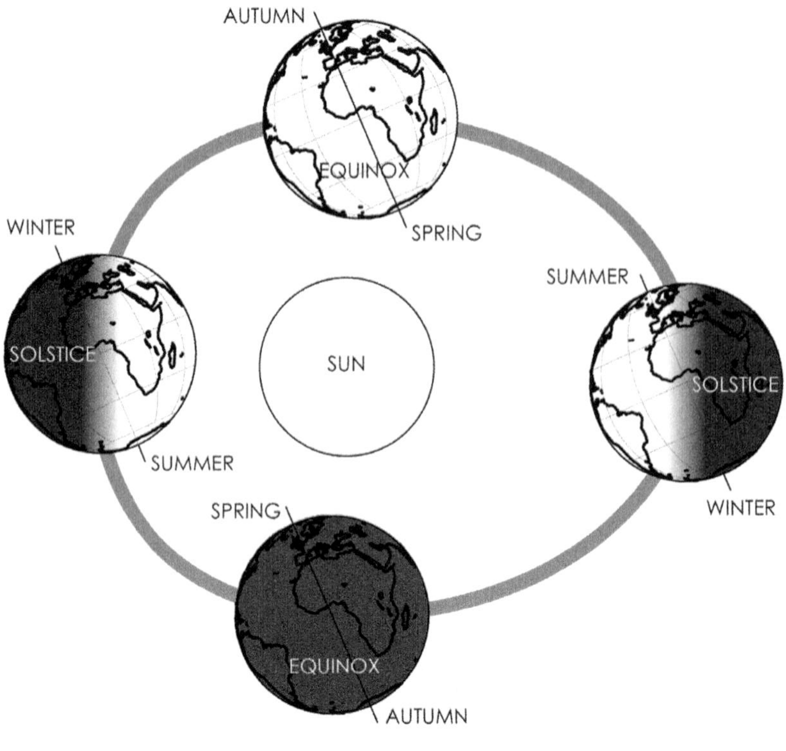

FIGURE 5.10 Illustration of the causes of Earth's seasons

changes in the length of day and the height of the sun in the sky are a result of the fact that Earth is tilted by about 23° – it does not spin around a vertical axis relative to its rotation around the sun. One way to understand that tilt is that the North Pole is not at the top of the Earth (Figure 5.10).

The tilt matters because its direction barely changes over the course of a human life-time, and so the axis is always pointing to the same star. Because Earth's tilt is practically constant, if you are standing in the right place you can use the Pole Star to find north, or the Southern Cross to find south, at any time of year. However, the constant tilt with respect to the stars also means that Earth's tilt is not constant with respect to the sun. In Figure 5.10, Earth's axis tilts toward the left. On about 20 June, the sun is also to the left of Earth, and so the North Pole is facing the sun, the sun is 23° above the horizon, and there is 24-hour daylight everywhere within 23° latitude of the North Pole (i.e., within the Arctic Circle, which is at about 67°N). At high noon, the sun is directly overhead at 23°N (the Tropic of Capricorn). Everywhere north of the Tropic of Capricorn the sun reaches its highest point above the horizon and it is the longest day of the year between the Tropic and the Arctic Circle. This date is the *boreal* (Northern Hemisphere) summer solstice (or the austral winter solstice from a Southern Hemisphere perspective).

On about 20 December, the sun is to the right of Earth, but Earth's axis is still tilted to the left, and so the North Pole is facing away from the sun. There is 24-hour night everywhere within the Arctic Circle, while everywhere else in the Northern Hemisphere experiences its shortest day on this date, and the sun's height above the horizon at high noon is the lowest for any time of the year. This date is the boreal winter solstice (or the austral summer solstice). At high noon the sun is directly overhead at 23°S (the Tropic of Cancer), and the Southern Hemisphere experiences its summer.

At the equinoxes (around 20 March and 22 September) the sun shines directly side-on to Earth – neither Pole points towards the sun. On these dates the sun is directly overhead at high noon at the equator, and everywhere on Earth has the same length of day and night. Those of us who live in the extratropics may use the position of the sun to identify north and south at high noon; within the Tropics the sun is sometimes to the north and sometimes to the south, depending on the time of year.

5.3.4 How do the seasons differ spatially?

As discussed in § 5.2.2, and illustrated in Figure 5.7, the seasons are most pronounced in high latitudes because of the strong contrast in the amount of sunlight between winter and summer at these latitudes. Within the tropics, the sun passes directly overhead twice per year. At the equator, these dates coincide with the equinoxes, but the length of day and the height of the sun in the sky does not change very much throughout the year. Therefore, 'summer' and 'winter' are not meaningful ways of defining the seasons here. Instead, close to the equator, the seasons are more meaningfully defined by rainfall than by temperature, although in some places very close to the equator even the rainfall remains nearly constant throughout the year.

Where the sun is directly overhead, Earth's surface is heated most intensely, and so, assuming there is a source of moisture, evaporation and convection (§ 4.2.2) are likely to be strong. There is a band of heavy rainfall that closely follows the sun as it moves north and south across the equator through the course of the year. This band of heavy rainfall is associated with the Inter-Tropical Convergence Zone (ITCZ). Strictly speaking, the ITCZ is defined in terms of winds rather than rainfall: it marks the line at which air from the Northern Hemisphere and the Southern Hemisphere converge into the low pressure that forms because of strong heating by the overhead sun. Nevertheless, the band of heavy rainfall closely follows the movement of the ITCZ, although in some longitudes (such as the south-eastern Pacific and the Middle East) there is too little moisture in the air for much rainfall to occur. The northern and southern limits of the ITCZ are modified by the positions of continents; it migrates furthest from the equator over land because the land heats up and cools down more quickly than the sea, and so it interacts with some of the main monsoon systems (§ 5.2.3). Between the northern and the southern limits, the ITCZ passes overhead twice per year and so many of these areas have two rainy seasons (as in much of East Africa, for example). Closer to the limits there is only the one rainy season (such as over continental South Asia and the Sahelian belt of West Africa).

Other areas with a single rainfall season include areas near the sub-tropics, such as the Mediterranean, Southern California, central Chile and the south-western parts of Australia and South Africa. These areas typically experience a wet winter and a hot dry summer because the subtropical deserts shift poleward in summer (§ 5.2.5) in response to the movement of the ITCZ. In the mid-latitudes, most areas have year-round rainfall, with large-scale rainfall dominating in the winter, and convective rainfall possibly becoming more important in the summer.

As well as geographical differences in the amplitudes of the seasons, there are also geographical differences in the lag in the seasons, and the lag can be strongly asymmetric. The lag tends to be longest in areas with coastal climates (§ 5.2.3), and is most extreme near the Poles. At the Poles, summer is delayed by only two or three weeks, but cooling continues throughout the winter until the sun starts to re-appear (March in the Northern Hemisphere, September in the Southern Hemisphere). As a result, near the North Pole, July is the warmest month, but occurs only five months after the coldest month, which is February. At the South Pole, because of the long distance from the sea, the lag is even more extreme: December is the warmest month, but occurs only four months after the coldest month, August.

5.3.5 How much does climate vary?

Just as latitude has a strong effect on changes in temperature throughout the year (§§ 5.2.3 and 5.3.4), it also dominates the variability in temperature from year-to-year (Figure 5.11 top). Temperature is most variable at high latitudes, partly because changes in wind direction can have a big influence on temperature especially in winter (§ 5.2.3), and partly because of sensitivity to changes in snow and ice cover (§ 4.2.7). Differences in temperature variability between inland and coastal areas are also evident: variability is stronger inland away from the dampening effect of the oceans (§ 5.2.3). Along the equator in the eastern and central Pacific Ocean temperature variability is relatively high for oceanic areas: the variability here is related to El Niño (Box 5.1).

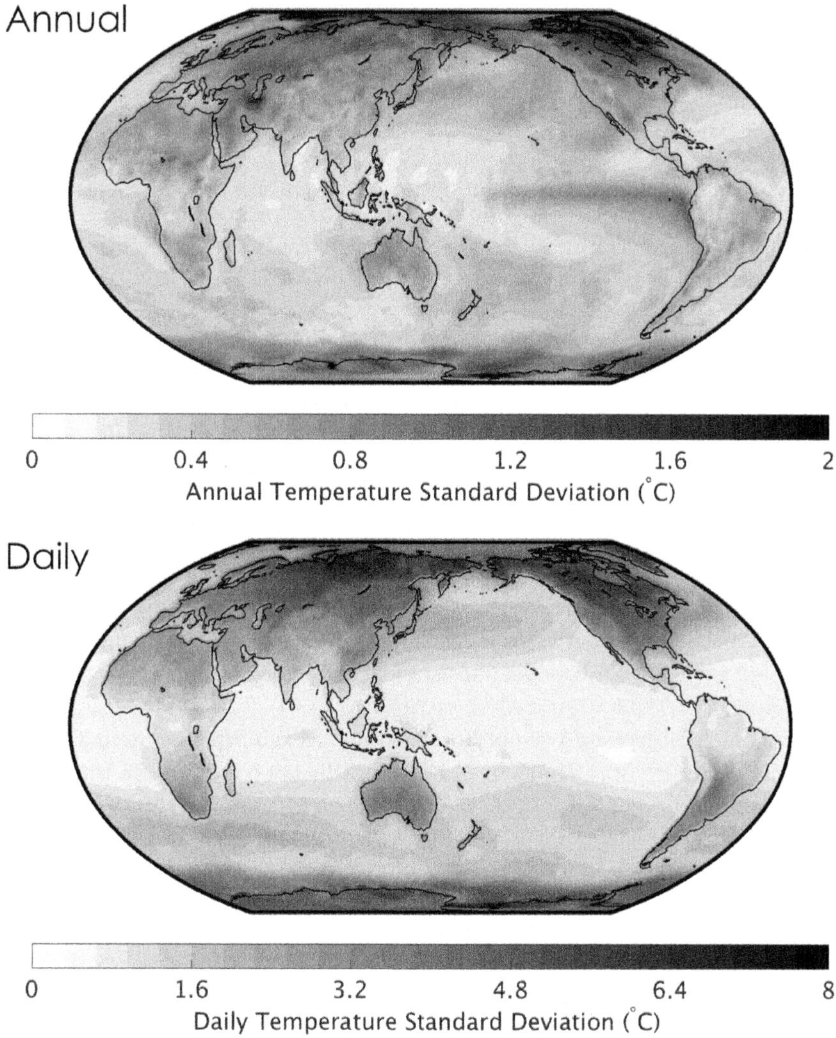

FIGURE 5.11 Year-to-year (top) and day-to-day (bottom) variability in temperature, as measured by the standard deviation, 1981–2010.
Data source: ECMWF Interim Reanalysis, for 1981–2010[1]

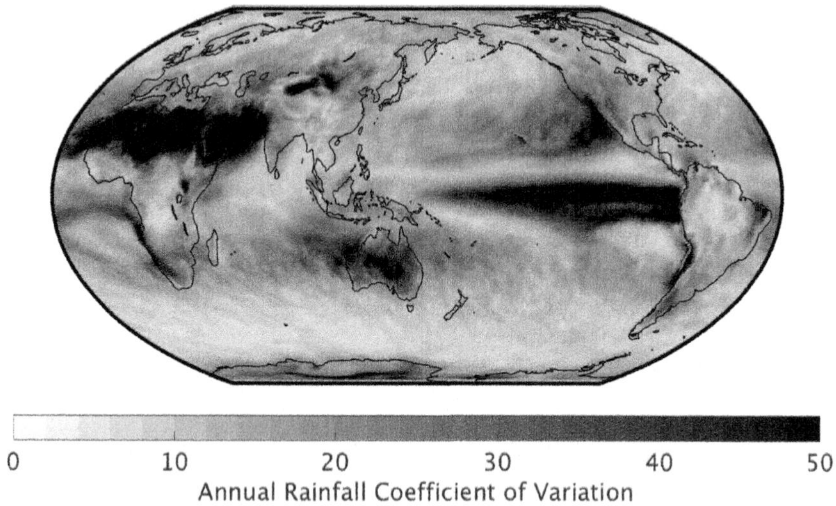

FIGURE 5.12 The coefficient of variation of annual rainfall, 1981–2010.
Data source: ECMWF Interim Reanalysis, for 1981–2010[1]

The spatial pattern of variability in temperature from day-to-day (Figure 5.11 bottom) is similar to that for variability from year-to-year (Figure 5.11 top), but temperatures vary from day-to-day at least four times more than they do from year-to-year over most of the globe outside of the tropics. Only over part of the tropical Pacific Ocean where El Niño occurs (Box 5.1) does temperature vary from year-to-year more than it does from day-to-day.

Comparing variability in rainfall at different locations is more complicated than for temperature. Because rainfall variability is affected by the average, the coefficient of variation (the standard deviation divided by the average) is used to control for this dependency (Figure 5.12). By this measure, variability is highest in the deserts, but generally is low in the mid-latitudes and humid tropics, and high in the subtropics.

Over much of the world, temperature and rainfall anomalies are related: if it is unusually wet or dry, it is frequently either unusually hot or cold at the same time. However, the nature of this relationship depends on location and season, and even time of day:

- Over the warm oceans, temperature increases with rainfall because of increased evaporation (i.e., temperature and rainfall are positively correlated).
- Over land, and especially in the tropics, daytime temperature decreases with increased rainfall because of increased cloudiness which blocks the sun's rays (i.e., temperature and rainfall are negatively correlated), but night-time temperature is positively correlated because of the blanketing effect of the clouds.
- Similarly, over the mid-latitude and subtropical land areas, summer daytime temperatures and rainfall are negatively correlated for the same reason.

- Near the poles, an increase in temperature in winter often occurs at the same time as an increase in snowfall because warmer air can hold more moisture (§ 4.2.3; i.e., temperature and rainfall are positively correlated).

All the correlations described above involve local relationships between rainfall and temperature, but they can be confounded if there are large-scale effects on the local climate, such as through El Niño.[9] During El Niño conditions, for example, maximum and minimum temperatures increase over much of the globe (Box 5.1), and where there is also an effect on rainfall the changes in temperature related to El Niño will be added to those associated with changes in rainfall. The resulting effect can become complicated.

5.4 Why does climate vary temporally?

Just as weather changes from day-to-day and week-to-week, so also does the climate change from month-to-month, from year-to-year, and beyond (Table 5.1). Some of this variability is as an effect of preceding weather and climate; these causes are understood as 'internal' to how climate works. Internal causes of variability include feedbacks or interactions in which the weather changes the Earth's surface, which in turn changes the weather. Asking which change comes first is a chicken and egg problem. Other causes of climate variability occur independently, perhaps because of human activity or because of major geophysical activity, or perhaps because of changes in how the Earth is heated by the sun. Independent changes are described as 'external'.

5.4.1 Internal causes of climate variability

Weather and climate are naturally variable, and primarily for the very reasons we have known about for thousands of years: 'the wind bloweth where it listeth' (*John 3:8*); 'the winds ... scatter clouds and rain' (*Sūrat al Mur'salāt 3*). In other words, the weather behaves like that sometimes. When we ask a question such as 'why is it foggy this morning?' we are usually satisfied with an answer that might describe moist winds from a warm ocean blowing inland where clear skies overnight have made the land-surface very cold. Such a response explains the fog in terms of the weather patterns, and any explanation of those weather patterns might refer in turn to air pressure patterns, jet streams, etc. – i.e., to other weather patterns. Climate is variable, in part because weather is so variable – the chaotic (as distinct from 'random') variability of weather means that the weather never repeats itself, just as it is impossible to get a pinball to follow exactly the same path. In some years, fog happens frequently because sometimes there are frequent cold nights and warm moist winds. And that is all the explanation there is. Sometimes. Sometimes, fog happens frequently because that warm ocean is warmer than it normally is, or those cold clear nights are colder and more frequent than normal, and there may be an explanation for those anomalies. We have already seen how important the Earth's surface is as a source of heat, of moisture, and because of how much sunlight it reflects (§§ 5.2.3, 5.2.4 and 5.3.2). Changes

TABLE 5.1 Timescales of weather and climate variability and trends, their causes and sources of uncertainty

Timescale of variability	*Timeframe*	*Primary drivers*	*Dominant sources of uncertainty for prediction*
Weather	Hours to 2 weeks	Preceding weather patterns	Initial conditions in the atmosphere (butterfly effect)
Sub-seasonal	1–4 weeks	Some large-scale weather patterns	Chaotic variability; model uncertainty
Seasonal	1–12 months	Earth's axial tilt drives basic seasonality; sea-surface temperature anomalies (especially those associated with El Niño – Southern Oscillation [ENSO]) sometimes affect the evolution of a particular season, primarily in the tropics	Chaotic variability; model uncertainty; initial conditions in the ocean
Multi-annual	1–10 years	Sea-surface and deeper ocean temperatures (e.g., ENSO and Atlantic variability); volcanic activity	Chaotic variability; initial conditions in the ocean; model uncertainty; timing of eruptions
Multi-decadal	10–30 years	Ocean temperatures and circulation; anthropogenic forcing	Model uncertainty; initial conditions in the ocean
Century	> 30 years	Anthropogenic forcing; solar variability	Greenhouse gas emissions; model uncertainty

in Earth's surface can therefore have an important effect on climate, and these surface changes may themselves be a result of unusual and/or prolonged weather. Ways in which Earth's surface and the air can interact to cause climate variability include via sea-surface temperatures, land temperature and soil moisture, and snow and ice cover.

5.4.1.1 Variability in Earth's surface

Sea-surface temperature anomalies affect evaporation and heating or cooling of the overlying air, and the effect can last for weeks, months or even longer, because it takes so much energy to change the temperature of water (§ 4.2.1). The effect on climate is strongest in the tropics[10] because changes in the amount of water evaporated from the sea are much larger in hot climates than in cold (§ 5.2.4).

BOX 5.1 EL NIÑO – SOUTHERN OSCILLATION (ENSO)

What is the El Niño – Southern Oscillation?

After the the seasons, the El Niño – Southern Oscillation (ENSO) is the most important example of climate variability. Although it occurs in the Pacific Ocean region, it can affect climate over many other parts of the globe. The ENSO consists of changes in the ocean (El Niño) and changes in the atmosphere (Southern Oscillation). An *El Niño* state occurs when the sea surface across much of the eastern and central Equatorial Pacific Ocean becomes unusually hot (Figure 5.13, top), by as much as 3 °C or even more in the strongest events. It is also possible for this part of the ocean to become colder than normal. Such cooling conditions are called *La Niña* (Figure 5.13, bottom). These warming and cooling periods typically last about 9–12 months, commencing around April or May, and lasting through to about March, and recur about every three to ten years.

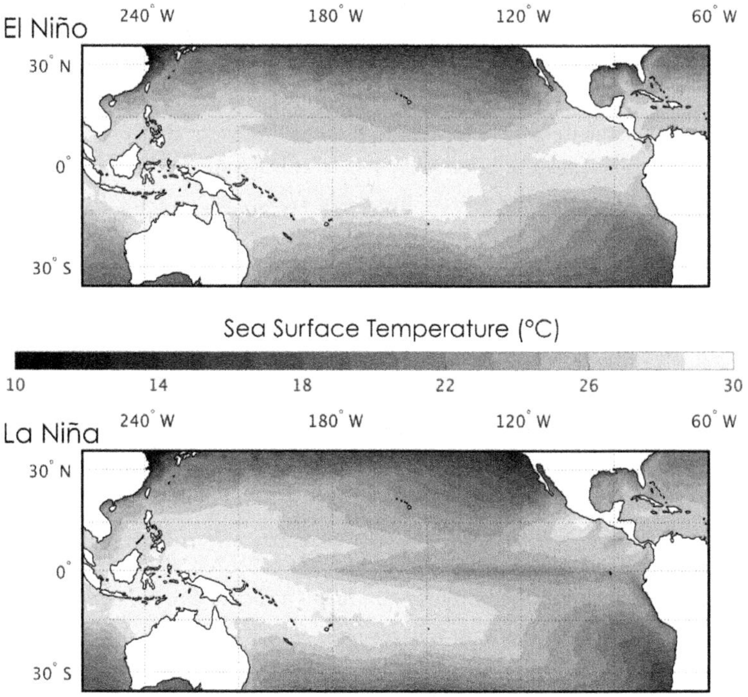

FIGURE 5.13 Sea-surface temperatures during (top) a strong El Niño event (Dec 1997–Feb 1998) and (bottom) a strong La Niña event (Dec 1998–Feb 1999). *Data source*: OISST2[11]

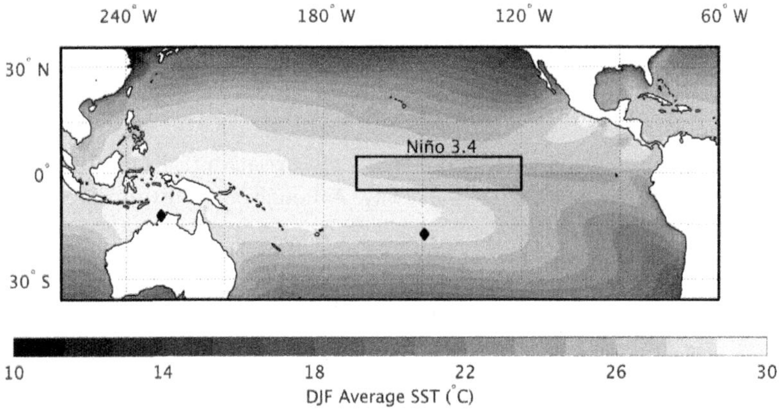

FIGURE 5.14 Average December–February sea-surface temperatures, 1982–2017. The black diamonds indicate the locations of Darwin and Tahiti, which are used in the calculation of the Southern Oscillation Index. *Data source*: OISST2[11]

The *Southern Oscillation* component of ENSO involves major changes in the position of high and low air pressure (§ 4.2.7) across the Pacific Ocean. The Oscillation measures air pressure differences between Darwin in Northern Australia, and at Tahiti in French Polynesia (at about 150°W in the central Pacific). In most years, air pressure is relatively low over Darwin, and high near Tahiti. This air pressure difference causes the Trade Winds, which blow from South America towards the western Pacific. (When describing the Pacific Ocean, it is easy to get confused about which side is east and which is west. The western Pacific is near Asia, which in most other contexts we understand as being the East; the eastern Pacific is near the Americas, which we understand as being the West.) Air pressure is low in the western Pacific because the sea is hot here (Figure 5.14), consistently exceeding about 28 °C, and the high humidity allows for heavy rainfall there (Figure 5.2). Further east, the cold ocean contributes to high atmospheric pressure and a drier climate.

The air and the sea are closely related across the equatorial Pacific in part because of an absence of the effects of land–sea contrasts, and so the Southern Oscillation and El Niño / La Niña vary in close relationship. During El Niño episodes, the warming in the eastern equatorial Pacific, where the sea is usually relatively cold (Figure 5.14), results in high temperatures extending across much of the ocean, thus weakening the air pressure difference. In strong El Niño conditions, sea-surface temperatures exceeding about 28 °C can extend across the entire width (Figure 5.13, top). During La Niña episodes, the cold eastern Pacific becomes even colder, and so the air pressure difference strengthens. La Niña thus looks much more like an enhancement of the average conditions (cf. Figures 5.12, bottom, and 5.13), whereas El Niño looks more like a change from the average (cf. Figures 5.13, top, and 5.14).

How often do El Niño and La Niña occur?

El Niño and La Niña conditions recur about every three to ten or more years, but their frequency and intensity vary inter-decadally[12] and inter-millennially. In the second half of the 19th and in the mid-20th century, for example, ENSO variability was relatively weak. The predictability of ENSO, and of climate variability in general, is poorer during these quiescent phases. Since the late-1960s ENSO has been more active, but we are unable to predict how long this phase will last.

Whenever El Niño and La Niña episodes occur, they typically dissipate in about March. However, occasionally, El Niño or La Niña can regenerate over the subsequent months, resulting in multi-year episodes. Since the mid-1800s, prolonged El Niño episodes have occurred every 20–60 years, the most recent occasions being in the early 1990s and mid-2010s. Prolonged La Niña episodes have occurred about twice as frequently, the latest being in the late 1990s.

How do El Niño and La Niña affect the climate?

El Niño and La Niña affect climate in a consistent way only in some parts of the world, and for only part of the year. The strongest effects are over the ocean, and only about 20% to 30% of land areas experience significant impacts on rainfall at least some part of the year. Different areas are affected in different seasons, and so rainfall over only about 15% to 25% of global land areas is affected in any particular season.[13] The strongest impacts are in the vicinity of the tropical Pacific Ocean, but rainfall in other parts of the globe can also be affected as large-scale climate patterns respond to a shift in the area of heavy rainfall between the eastern and central Pacific described above. The International Research Institute for Climate and Society has developed simple web-based tools that can be used to explore the relationship of ENSO to rainfall and temperature across the globe[i] or more specifically in Africa.[ii,14] Southern African rainfall is strongly impacted by ENSO with drought disasters associated with El Niño years[15] and high malaria incidence anomalies with La Niña years (see Case Study 5.2).

Impacts of El Niño and La Niña on air temperature are less well-studied than for rainfall. The warming of the equatorial Pacific Ocean during El Niño episodes does contribute to a notable increase in global average temperature and to warming over much of the tropics. The effects on air temperature are largely a combination of the increased heating from the warmer oceans, which extends into the Indian Ocean because of changes in wind, and of changes in cloudiness associated with shifts in rainfall patterns. Epidemics of malaria in highland regions of Ethiopia and Colombia have been notably associated with warmer El Niño years.[21]

CASE STUDY 5.2 IMPACT OF RAINFALL AND THE EL NIÑO – SOUTHERN OSCILLATION ON MALARIA IN BOTSWANA

Madeleine C. Thomson, IRI, Columbia University, New York, USA

Botswana has made substantial progress towards eliminating malaria over the last decade. Malaria cases reduced from 17,886 (0.97% prevalence) in 2008

TABLE 5.2 National de-trended (standardized) confirmed malaria cases (1982–2003) in Botswana during the malaria season (January–May)[17] and their relationship to December–February rainfall (estimated from Merged Analysis of Precipitation from the Climate Prediction Center [CMAP]). The table is ordered from low to high malaria incidence years. El Niño events (defined using the Oceanic Niño Index[iii]; see Box 6.3) in the months preceding and during the malaria season are mostly associated with low malaria anomalies while La Niña events are largely associated with high malaria years.[17,19] 1988 was affected by both El Niño and La Niña events

Year (ordered by malaria incidence anomaly – lowest to highest)	*Malaria incidence anomalies*	*Total DJF rainfall (in mm)*	*ENSO phase*
1992	−1.88	153	El Niño
1982	−1.57	162	
2002	−1.45	158	
1987	−1.18	170	El Niño
1983	−1.12	163	El Niño
2003	−1.11	200	El Niño
1995	−0.72	171	El Niño
1984	−0.62	174	
1991	−0.24	268	
1985	−0.14	222	La Niña
2001	−0.14	208	La Niña
1990	0.02	219	
1998	0.10	197	El Niño
1994	0.32	300	
1986	0.50	214	
2000	0.80	439	La Niña
1999	0.87	240	La Niña
1989	1.25	354	La Niña
1997	1.33	320	
1993	1.50	223	
1996	1.54	342	
1988	1.95	330	El Niño, La Niña

Abbreviations: DJF, December, January, February; ENSO, El Niño – Southern Oscillation.

to 311 (0.01% prevalence) cases in 2012.[16] Prior to recent investments in malaria control and elimination, year-to-year variability in malaria incidence was pronounced and could largely be explained by variations in rainfall during the December–February season once longer term trends (likely associated with control and treatment measures) had been removed.[17] Rainfall in much of Southern Africa closely follows sea-surface temperatures in the eastern equatorial Pacific, and major droughts or wet years are significantly associated with El Niño and La Niña (Box 5.1).[18] An analysis of malaria and climate 1982–2003 (Table 5.2) indicates the strength of the relationship between malaria, rainfall and sea-surface temperatures prior to the implementation of the elimination strategy. This relationship forms the basis for seasonal climate forecasts to provide early warning of malaria epidemics.[19] Incorporating climate into malaria impact assessment is necessary to avoid over or under-estimation of the impact of malaria interventions.[20] In March 2017 the Ministry of Health notified the public that the country was experiencing an unusually high level of malaria transmission following a period of heavy rains. Distinguishing between climate and other drivers (such as control factors) of year-to-year variations in malaria transmission continues to be of critical importance to malaria control managers.

Sometimes a change in sea-surface temperature is reinforced by changes in the weather that it causes. The air can change sea temperatures by various mechanisms: winds drive ocean currents, mix the warmer surface layers with colder subsurface water, and affect the rate of evaporation (the evaporation rate increases with strong winds), while evaporation and rain can have a direct effect on ocean temperature and an indirect one by affecting salinity. Salinity is important because it affects the density of water, which in turn drives ocean currents in a similar way to water temperature.[10] There are many examples of how the sea and the air affect each other to create changes in climate that operate on a wide-range of timescales. At some point, the interaction either breaks down or reverses, creating an 'oscillation' in which a climate anomaly develops, persists for a while, and reverses as a new mechanism takes over. The best known of these oscillations is the ENSO (Box 5.1). The ENSO is a Pacific Ocean phenomenon, but can affect climate around the world. The Atlantic Equatorial Mode and the Indian Ocean Dipole are analogous phenomena in the other tropical oceans, and which can have important regional impacts on climate.[10] Other examples of important climate oscillations are described in Box 5.2. Some of these oscillations can operate on timescales lasting many years.

Climate variability can also be caused by changes in the land surface by affecting how much sunlight is reflected (§ 4.2.7), especially in the presence of snow, and by affecting the amount of evaporation into, and heating of, the air (§§ 4.2.7 and 5.2.4). Examples include the influence of the Himalayan snow pack on the

BOX 5.2 CLIMATE OSCILLATIONS

Ángel G. Muñoz, IRI, Columbia University, New York, USA

While ENSO is the most important mode of climate variations at the global scale and tends to have the greatest impact in the tropics, other important climate oscillations exist with more localized impacts.

North Atlantic Oscillation

The North Atlantic Oscillation (NAO) is a large-scale pattern of natural climate variability characterized by a seesaw difference in air pressure between the Azores and Iceland. The Oscillation has important effects on rainfall and temperatures across the eastern United States, and much of Europe, extending as far as the Middle East. Successfully predicting the Oscillation is key to making accurate seasonal forecasts of climate in these areas.[10] Until recently, models were unable to predict the Oscillation, but recent improvements in model resolution and initialization schemes seem to be enhancing the skill at seasonal timescales.

Madden-Julian Oscillation

The Madden-Julian Oscillation (MJO) involves an area of strong convection and heavy rainfall that moves eastward through the global tropics. The eastward movement occurs at varying speed, and can take from 30 to 60 days to circle all the way around the Earth. For areas in the tropics, the Oscillation can bring a series of wet and dry spells. Some effects outside of the tropics can also be felt. Recent advances in numerical modeling of the MJO are providing promise for forecasting weather at timescales of one week to one month, mostly because of improved modelling of rainfall processes.

South Asian monsoon, and the importance of soil moisture deficits as precursors to heatwaves.[10]

5.4.2 External causes of climate variability

Some changes to the Earth's surface occur independently, perhaps because of human activity or because of major geophysical activity. However, changes in Earth's surface are not the only external cause of climate variability: the amount of energy received from the sun can vary; and although Earth's surface provides heat and moisture to the air, how much of that heat and moisture the air retains can change because of the composition of the air. Such changes in the composition of the air can be natural or human-induced.

5.4.2.1 Volcanoes

The impact of a volcanic eruption on climate depends on much more than the violence of the eruption: the direction, location and chemistry of the eruptions are all important. Some volcanic eruptions, such as Mount St Helens in 1980, explode laterally, and so the dust and ash stay near Earth's surface where they can be washed out of the atmosphere by rain in a few days. Other eruptions, like Mount Pinatubo in 1991, explode vertically, and so the emissions can get high into the atmosphere where they can remain for months or even years, and cool the Earth by blocking sunlight from reaching the surface. These vertical eruptions have longer-lasting impacts than more lateral eruptions. For example, Mount Pinatubo's eruption cooled the Northern Hemisphere climate by about 0.6 °C as well as contributing to widespread decreases in rainfall in the tropics.[22] These impacts are considerably more than that of a major El Niño (Box 5.1). The very largest eruptions can have devastating impacts, and can change the climate for millennia if the cooling is sufficient to cause widespread snowfall and freezing.

Because of the direction of high altitude winds, particles from volcanic eruptions that do get above about 15 km are slowly transported towards the Poles (in the same way that ozone is transported towards the Poles; Box 4.4). Therefore, eruptions from volcanoes near the equator may have a more widespread impact than high latitude eruptions. The cooling effect is strongest when the eruption emits large volumes of sulphates, since these form aerosols that are highly effective in blocking sunlight. Indeed, sulphate aerosols from a series of relatively small near-equatorial eruptions are partly responsible for a slow-down in the global warming trend[23] over the first approximately 15 years of the 21st century. In addition to sulphates, volcanoes do emit CO_2, which contributes to the greenhouse effect (see § 9.3), but the average annual emission from volcanoes is less than 1% of human emissions, and so this greenhouse effect is negligible.

5.4.2.2 Solar variability

The amount of energy emitted by the sun varies on a fairly strict cycle of about 11 years, but the strength of these cycles itself changes on a less-predictable timeframe. One manifestation of these changes is the appearance of dark spots on the sun, which represent areas of stronger activity. There are reliable records of these sunspots extending back hundreds of years. During a quiescent period in the 1800s, decreased solar activity contributed to the development of the Little Ice Age, during which European winters were bitterly cold. Solar activity picked up in the mid-1800s, about the same time as the industrial revolution was causing an increase in greenhouse gas concentrations. Solar activity seems to have reached a peak in the 1990s, and has been decreasing since.

At timescales of thousands of years, changes in Earth's orbit affect how far the Earth is from the sun at different times of year, as well as by how much the amount of sunlight changes at different latitudes over the course of a year. These orbital changes can be projected forward and backwards in time many hundreds of thousands of years because they are based on the gravitational effect of the planets and the moon, whose movements are known in detail. The changes match exceptionally well with the advance and retreat of Ice Age conditions.

5.4.2.3 Atmospheric composition

The sun heats the Earth and the Earth heats the air. However, just as most of the air is transparent to most of the sun's radiation, so also the air is transparent to some of Earth's emitted radiation. We have already seen that ozone can absorb certain types of radiation that other gases cannot (§§ 4.2.6 and 4.2.7); in fact, each gas is able to absorb different types of radiation. Gases that absorb Earth's radiation are called greenhouse gases. On Earth, the main greenhouse gases are water vapour, carbon dioxide and methane. The most effective greenhouse gases are those that absorb radiation that other gases do not: similarly, a small board that can block a hole in the window will likely insulate your house more effectively than adding a second layer of loft insulation over tens of square metres. Because greenhouse gases can plug these metaphorical holes, they can be important even when their concentrations in the air are low. If Earth had no greenhouse gases it would be more than 30 °C colder than it is.

Changes in the amount of greenhouse gases in the air will affect its temperature, and thereby can alter the climate. Water vapour is easily the most abundant greenhouse gas on Earth, and its effects may even be sensible from day-to-day: cloudy nights are generally so much warmer than clear nights, for example (see further discussion in § 5.3.1). Because evaporation generally increases as temperatures rise, water vapour can serve to enhance the warming caused by other greenhouse gases. It can therefore be misleading to quote the warming potential caused by an increase in the concentration of a greenhouse gas in isolation. There are many complicating feedbacks.

After water vapour, carbon dioxide (CO_2) is the next most abundant greenhouse gas. It occurs naturally in the atmosphere, and natural variability is clearly evident in its annual cycle through the effects of plant growth. There is slightly over twice as much land in the Northern Hemisphere than in the Southern, and that difference is even greater when one considers only the latitudes with distinct growing seasons. As a result of this inequality in land distribution, concentrations vary by about 1.5% over a year, peaking at the end of winter before the Northern Hemisphere spring when plant growth starts to absorb the gas, and reaching a minimum at the end of summer when the leaves begin to fall. That 1.5% fluctuation is too small to have a significant effect on climate, but CO_2 concentration has increased by about 40% over about the last 250 years because of human activities, primarily through the burning of fossil fuels. The current rate of increase over about three years is equivalent to the increase between autumn and spring as a result of the annual cycle mentioned above.

Methane is another important greenhouse gas, and has increased by about 150% over the same period as CO_2. The methane increase is primarily a result of livestock farming. There is concern that melting of the permafrost in high latitudes because of global warming will release large amounts of additional methane into the air, thus enhancing the greenhouse effect further.

5.5 Conclusions

Spatial and temporal variations of temperature are much simpler than those of rainfall at virtually all scales. It is important to understand these scales of variability in space and time to obtain some idea of the necessary resolution of data for any analyses of climate–health relationships. Such an understanding can also contribute to an awareness of some of the limitations of these analyses given the constraints that data availability may impose. The following chapter provides an introduction to the nature, availability and limitations of climate data.

Notes

i https://iridl.ldeo.columbia.edu/maproom/ENSO/Impacts.html.
ii http://iridl.ldeo.columbia.edu/maproom/Health/Regional/Africa/Malaria/ENSO_Prob/ENSO_Prob_Precip.html.
iii http://origin.cpc.ncep.noaa.gov/products/analysis_monitoring/ensostuff/ONI_v5.php.

References

1 Dee, D. P. *et al.* The ERA-Interim reanalysis: configuration and performance of the data assimilation system. *Quarterly Journal of the Royal Meteorological Society* **137**, 553–597 (2011).
2 Aregawi, M. *et al.* Measure of trends in malaria cases and deaths at hospitals, and the effect of antimalarial interventions, 2001–2011, Ethiopia. *PLOS One* **9**, e106359 (2014).
3 Lyon, B., Dinku, T., Raman, A. & Thomson, M. C. Temperature suitability for malaria climbing the Ethiopian highlands. *Environmental Research Letters* **12**, 064015 (2017).
4 Moore, G. W. K. & Semple, J. L. Weather and death on Mount Everest: an analysis of the into thin air storm. *Bulletin of the American Meteorological Society* **87**, 465–480 (2006).
5 Shukla, J., Nobre, C. & Sellers, P. Amazon deforestation and climate change. *Science (Washington)* **247**, 1322–1325 (1990).
6 Bierlein, J. F. *Parallel Myths*. 368pp (Random House Publishing Group, New York, 2010).
7 Lebel, T., Delclaux, F., Le Barbe, L. & Polcer, J. From GCM scales to hydrological scales: rainfall variability in West Africa. *Stochastic Environmental Research and Risk Assessment* **14**, 275–295 (2000).

8 Ma, Y. & Guttorp, P. Estimating daily mean temperature from synoptic climate observations. *International Journal of Climatology* **33**, 1264–1269 (2013).
9 Omumbo, J., Lyon, B., Waweru, S. M., Connor, S. & Thomson, M. C. Raised temperatures over the Kericho tea estates: revisiting the climate in the East African highlands malaria debate. *Malaria Journal* **10**, 12, doi:10.1186/1475-2875-10-12 (2011).
10 Mason, S. J. in *Climate Information for Public Health Action* (eds. M.C. Thomson & Mason S.J.), Ch. 8 (Routledge, London, 2018).
11 Reynolds, R. W. *et al.* Daily high-resolution-blended analyses for sea surface temperature. *Journal of Climate* **20**, 5473–5496 (2007).
12 Trenberth, K. E. General characteristics of El Nino-southern oscillation in *Teleconnections Linking Worldwide Climate Anomalies* (eds. Glantz, M.H., Katz, R.W. & Nicholls, N.), 13–42 (Cambridge University Press, Cambridge, 1991).
13 Mason, S. J. & Goddard, L. Probabilistic precipitation anomalies associated with ENSO. *Bulletin of the American Meteorological Society* **82**, 619–638 (2001).
14 Thomson, M. C., Muñoz, A. G., Cousin, R. & Shumake-Guillemot, J. Climate drivers of vector-borne diseases in africa and their relevance to control programmes in Special Issue: *Vector-borne Diseases under Climate Change Conditions in Africa. Infectious Diseases of Poverty* (in press).
15 Thomson, M. C., Abayomi, K., Barnston, A. G., Levy, M. & Dilley, M. El Niño and drought in southern Africa. *Lancet* **361**, 437–438 (2003).
16 Simon, C. *et al.* Malaria control in Botswana, 2008–2012: the path towards elimination. *Malaria Journal 12*, 458 (2013).
17 Thomson, M. C., Mason, S. J., Phindela, T. & Connor, S. J. Use of rainfall and sea surface temperature monitoring for malaria early warning in Botswana. *American Journal of Tropical Medicine and Hygiene* **73**, 214–221 (2005).
18 Mason, S. J. El Niño, climate change, and Southern African climate. *Environmetrics* **12**, 327–345 (2001).
19 Thomson, M. C. *et al.* Malaria early warnings based on seasonal climate forecasts from multi-model ensembles. *Nature* **439**, 576–579 (2006).
20 Thomson, M. C. *et al.* Using rainfall and temperature data in the evaluation of national malaria control programs in Africa. *American Journal of Tropical Medicine and Hygeine* **97**, 32–45, doi:10.4269/ajtmh.16-0696 (2017).
21 Siraj, A. S. *et al.* Altitudinal changes in malaria incidence in highlands of Ethiopia and Colombia. *Science* **343**, 1154–1158, doi:10.1126/science.1244325 (2014).
22 Winter, A. *et al.* Persistent drying in the tropics linked to natural forcing. *Nature Communications 6*, 7627 (2015).
23 Fyfe, J. C., Gillett, N. P. & Zwiers, F. W. Overestimated global warming over the past 20 years. *Nature Climate Change* **3**, 767–769 (2013).

6

CLIMATE DATA

The past and present

Simon J. Mason, Pietro Ceccato and Chris D. Hewitt
Contributors: Theodore L. Allen, Tufa Dinku, Andrew Kruczkiewicz, Asher B. Siebert, Michelle Stanton and Madeleine C. Thomson

> I'll show you how t' observe a strange event.
>
> Timon of Athens *by William Shakespeare*

6.1 Introduction

Historical and current weather and climate observations and monitoring products are essential for health risk assessment, planning and for the development of climate-informed early warning systems (see § 3.4). They are also important in assessing the efficacy of climate-sensitive interventions.[1] Historical climate data are also needed as a baseline for assessing any changes in climate, for developing and evaluating climate models used in predictions and projections, and for providing the initial conditions for climate predictions[2] (see §§ 8.2.2 and 9.4). However, only a few weather observations can be used for climate work because the objectives in taking the observations differ from those for climate observations. For weather forecasts, the best possible observations are required to make accurate predictions; if the observations can be improved in any way – perhaps by moving the station, or using more accurate instruments – then such improvements offer an immediate advantage. In contrast, consistency of observation is a key consideration for climate work. Climate scientists want to compare observations (is this year hotter than last year?). If an instrument is changed or moved to a different location, then it makes it difficult to make comparisons.

In this chapter we describe how climate variables are measured (e.g., ground observation, remote sensing or modelled data), how the data are collected and how they are shared. We discuss the advantages and disadvantages of different data sources and highlight the potential advantages of climate data made available through National Meteorological and Hydrological Services. The use of observed climate monitoring products in the development of early warning systems is highlighted.

6.2 How are global weather and climate data produced and shared?

Accurate weather and climate monitoring and forecasting requires a massive international coordination of data collection and sharing (see, for example, §7.4). In 1963, the World Meteorological Organization (WMO) implemented the World Weather Watch as a system for facilitating this coordination.[3] The Watch involves three components, one to take the observations (the *Global Observing System*), another to share the observations and resulting information products (the *Global Telecommunication System*) and a third to generate those information products from the data (the *Global Data Processing and Forecasting System*).

6.2.1 Global Observing System

The Global Observing System (GOS) provides observations of the air and the ocean surface from surface weather stations and ships, ocean buoys, weather balloons, satellites and other sources. The observations are collected by National Meteorological and Hydrological Services (NMHSs) and various satellite agencies and consortia, all of which collect the observations according to specified standards. Some of the observations (the 'synoptic' observations) are collected at coordinated times so that these can be used to initialize weather prediction models (see §7.4).

Although most of the observations are used for weather prediction, a few observations are specifically for climate work. These observations form part of the *Global Climate Observing System* (GCOS),[4–6] which provides the long-term observations required for monitoring, research and prediction of climate change and variability and their impacts, such as for the Intergovernmental Panel on Climate Change (IPCC) Assessment Reports[7] (see § 9.2). Countries are supposed to report monthly statistics from these stations to global climate monitoring centres in Germany and Japan. Although the frequency of on-time reporting has been historically poor from many areas outside of the northern mid-latitudes, it has improved dramatically since 2012 with the formation of the Global Framework for Climate Services (GFCS)[8] (see Box 1.3). These GCOS observations form the basis for some of the freely available and most commonly used global climate datasets (§ 6.4). However, there are many additional observations that are not routinely shared with regional and global data centres. These additional observations may be available only from NMHSs and other national centres. Because there are much more data available at the national level than at the global level, countries have the possibility of producing datasets that are of higher quality than global products (Case Study 6.1).

CASE STUDY 6.1 ENHANCING NATIONAL CLIMATE SERVICES (ENACTS) DATA PRODUCTS

Asher B. Siebert and Tufa Dinku, IRI, Columbia University, New York, USA

The recently developed Enhancing National Climate Services (ENACTS) merged station-proxy temperature and rainfall data products[9] have been found to be of high value for climate analysis, especially in regions with sharp climatic contrasts over a small geographic domain. ENACTS products use all available meteorological station data from participating countries. For rainfall, the station data are blended with the most appropriate global satellite rainfall products (such as CHIRPS[i] or TAMSAT[ii]) using an interpolation algorithm to construct a temporally and spatially complete gridded product. For temperature, the station network is blended with reanalysis (such as the Japanese 55-year Reanalysis[10]) and digital elevation model data.

One country that has sharp spatial climatic contrasts that has benefited from ENACTS is Rwanda. Situated in the East African highlands, Rwanda is a small, densely populated country, whose citizens depend largely on rain-fed agriculture. Despite its small size (~26,000 km^2) the landscape ranges from relatively lush savannah grassland to cloud rainforests. Elevation ranges from just under 1000 m to almost 4500 m above sea-level.

Within the last several years, high resolution ENACTS data products (with national coverage and ~4 km spatial resolution) and services have been established by Meteo Rwanda with the support of the International Research Institute for Climate and Society. The rainfall products are available from the early 1980s to the present, and are particularly valuable as there was a precipitous decline in meteorological observations from the early/mid 1990s until almost 2010 because of the Rwandan genocide, civil war and aftermath (See Figure 6.1). The ENACTS products fill the data gaps with satellite data adjusted using historical and more recent stations observations. The resurgence in the number of meteorological stations in ENACTS data post-2009 is not reflected in global datasets such as the Global Precipitation Climatology Centre (GPCC) (see Figure 6.1) re-iterating the point that more data are available locally than ever make their way into the global archives.

Because Rwanda sits just a few degrees off the equator, temperature variability in time is quite low and temperature variability in space is controlled largely by elevation (§ 5.2.1). Rwanda experiences a bimodal pattern of rainfall with rainy seasons in March–May and September–December, a partially dry period in January–February and very dry period from June–August. For rainfall, the higher elevations, which are the wettest regions of this small country in the northwest and southwest, can receive in excess of 2 m of rainfall per year on average, while the driest regions in the lowland east and south may receive less than 700 mm of average annual rainfall.

The very high resolution ENACTS data products are able to capture many of the nuances of Rwanda's climate in ways that coarser datasets fail to do and can thus provide valuable information for farmers and decision-makers in multiple sectors. Furthermore, in some preliminary studies, ENACTS products are shown to be a more skilful tool for calibration of seasonal forecasts than other comparably high-resolution global rainfall datasets (e.g., CHIRPS). However, users should not automatically assume that higher resolution data gives a clearer picture of the real spatial and temporal climate variability without careful evaluation and verification.

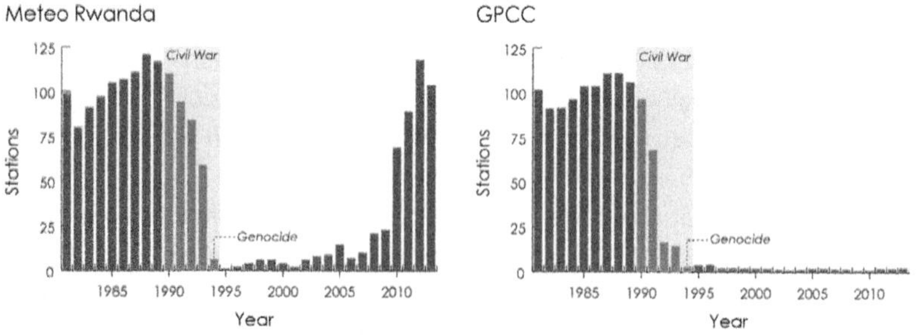

FIGURE 6.1 Number of functioning meteorological stations by year in Rwanda that provide data in a) ENACTS and b) GPCC rainfall products

6.2.2 Global Telecommunication System

The Global Telecommunication System (GTS) enables the communication of the observations to a network of centres that produce global or regional weather analyses and forecasts (§§ 6.3.3 and 7.4.2). This coordinated exchange of observations and model data is critical for the timely and accurate generation of warnings, and is promoted through a set of international agreements on data exchange (Box 6.1). Those forecasts are then transmitted back to the NMHSs for further downscaling and interpretation. Since 2003, the WMO has begun to upgrade the GTS to take advantage of new technologies (see Box 10.1) and to facilitate exchange of information beyond the network of Meteorological Services.

6.2.3 Global Data Processing and Forecasting System

The Global Data Processing and Forecasting System (GDPFS) performs quality control on the collected observed data, and uses the observations to produce analyses (§§ 6.3.3 and 7.4.2), monitoring products, and forecasts, which are then made available to the NMHSs. This system involves a network of national, regional and international centres that not only makes weather forecasting possible, but which

BOX 6.1 DATA-SHARING POLICIES

Agreements for the exchange of essential weather information have been in place since 1947 when the Convention of the WMO was signed. Promotion of the collection and exchange of meteorological data is effectively listed as the primary role of the WMO.[11] In practice, international data exchange has been sub-optimal, in part because of lack of clarity about what 'essential' means, and what the rules for non-essential data are. To address these issues, WMO has passed a series of Resolutions to clarify definitions and the rules over the exchange of non-essential data.[12] There have been three important such Resolutions[iii]:

- WMO Resolution 40 (1995; *WMO policy and practice for the exchange of meteorological and related data and products including guidelines on relationships in commercial meteorological activities*) reaffirmed the commitment to worldwide co-operation in the establishment of observing networks and the free and unrestricted exchange of meteorological and related information for non-commercial purposes.[13] Although the data are supposed to be 'freely' available, provision is made for costs incurred in distributing the data.
- WMO Resolution 25 (1999; *Exchange of hydrological data and products*) extended the provisions of Resolution 40 to include hydrological data.
- WMO Resolution 60 (2015; *WMO policy for the international exchange of climate data and products to support the implementation of the GFCS*) extends the definition of climate data given in Resolution 40 to include all 'GFCS relevant data and products' and reaffirms the commitment to their free and unrestricted exchange.

What do all these Resolutions mean for people working in public health? In theory, it should be possible to access relevant global, regional and national data and information products that are essential for serving the public good[8] at no more than the cost of redistribution. In practice, however, the required data may: a) not exist; b) be difficult to use and interpret because of missing values and other quality issues; c) still come at a significant cost because many NMHSs are provided with insufficient funds to collect and manage the data in the first place; and d) be provided with reluctance or not at all because of concerns over how the data may be used (or mis-used).

has enabled sustainable provision of information products and services for hazardous conditions, such as tropical cyclones, storm surges and the provision of operational meteorological services in the case of nuclear and other technological accidents, wild fires and volcanic eruptions.

6.2.4 Global Atmospheric Watch

In addition to the World Weather Watch, the WMO also coordinates the Global Atmosphere Watch (GAW) Programme,[14] which collects information on air quality and composition (§§ 4.2.6 and 4.2.7.3). The primary foci of the GAW include greenhouses gases, the ozone layer, reactive gases[15] and other air pollutants, and ultraviolet radiation. The GAW programme has played a key role in providing data in support of initiatives such as the Montreal Protocol on Substances that Deplete the Ozone Layer (Box 4.4) and the IPCC (§ 9.2).

6.3 What types of meteorological data are available?

Unfortunately, there is no single climate dataset that is likely to address all public health purposes. All the datasets have some drawbacks, and choices need to be made between length of record, availability in near-real time, numbers of missing values, accuracy, consistency, temporal and spatial resolution, etc. Strengths and weaknesses of the different datasets are best described by classifying the data based on how they are generated. Some data are obtained by direct measurement at a weather or climate station (e.g., temperature by thermometer), others by proxy measurement (e.g., tree ring width, isotope analyses or by remote sensing estimations), while others are generated theoretically using models (e.g., analyses and predictions[16]).

6.3.1 Direct measurements from climate stations

6.3.1.1 In situ station data

Some climate stations have records extending back more than 100 years, although most have less than 50 years. Monthly accumulations or averages are by far the most widely available resolution, but daily and sub-daily data may be obtainable. Although extensive use can be made of the monthly data, information about individual severe weather events may get lost or severely dampened, and so some significant events of relevance to public health may not be well-represented at this resolution. In many countries, much of the daily and sub-daily data remain as hand-written or graphical records. There are ongoing data rescue programmes to digitize these data to make them available for product generation and analysis (Box 6.2).

There are challenges when using historical weather records because the observations may not have been collected consistently. The weather station may have been moved or the instruments on it may have been replaced or upgraded so that some of the measured changes over time have nothing to do with climate change or variability. It may be possible to correct for these inhomogeneities, especially if information about changes in how the data were collected (so-called meta-data) is available.

Even when the data are consistent, there are likely to be some missing values, especially at daily and finer resolutions, and so some degree of quality control

BOX 6.2 DATA RESCUE

Theodore L. Allen, IRI, Columbia University, New York, USA

In many places modern technology automates rain gauge observations to record rainfall continuously and saves the information directly onto an electronic medium for archival and analysis purposes. However, in the absence of modern technology, automated rainfall observations are recorded directly onto paper by hand and are not automatically archived in a digital format. Data rescue projects are designed to transfer archived paper-based meteorological observations onto an electronic medium so that the information is readily available for scientific analysis using computational software programs.[17] Because analysing long-term meteorological observations from paper records is a very inefficient and time consuming task, electronic digitization software has been developed to automatically convert scanned paper records into a computer ready electronic format. Digitization software reads strip chart paper data, which is a common recording medium for automated weather stations that measure rainfall, temperature, wind speed, wind direction and atmospheric surface pressure. Data rescue projects still exist for non-automated observations, which are taken by hand, but they require an arduous task of manually typing the written observations onto a computer. The Atmospheric Circulation Reconstructions over the Earth (ACRE) initiative undertakes and facilitates most global data rescue projects with the aim of converting historic paper-based data into modern gridded datasets commonly used for climate models. Results from ACRE-led projects provide usable meteorological data that were at one time locked in a paper format.

is almost always required before using station data in health analyses.[18] It may be possible to estimate these missing values from neighbouring stations or from remotely sensed data, but for some analyses it may be sufficient to know only whether an extreme event occurred without details of its magnitude. Unfortunately, missing values disproportionately occur during severe weather and climate events when instrumental failures and breaks in communication are most likely to happen. Consequently, lower quality data are also most likely to occur in developing countries where climate impacts on the population may be particularly severe.

6.3.1.2 Gridded station data

Meteorological stations are not always representative of a location where data are needed by health specialists. What is to be done if there is no station in the location of interest? One option is to use interpolated station data. Some of the most commonly used climate datasets are constructed from station data interpolated to a regular grid (§ 6.4). These gridded datasets are ideal for large-scale (sub-continental) analyses, and for some country-scale analyses. However, although some quality control checks and corrections have been performed on the station data, serious inhomogeneities may remain in data-poor regions and periods. In most areas, the number of stations used in these datasets peaks in the 1950s–1970s, and drops off considerably before and after then. The changes in the locations and numbers of available stations can have serious implications for the quality of the interpolations (Case Study 6.1). In most cases, information about the number of stations in the vicinity of each gridpoint should be available, and this information needs to be checked before using the data, particularly for rainfall because of its localized nature (§ 5.2.5.2). The number of stations used in the global and regional gridded datasets are generally far fewer than are available nationally, and so more detailed and more accurate data may be available for national and sub-national analyses.

6.3.1.3 Index datasets

Some station data are used to measure specific climate phenomena rather than local climate at a distinct location. Data from a set of carefully selected stations may be combined into an index to represent such phenomena. Examples include the Southern Oscillation (Boxes 5.1 and 6.3) and the All-India Monsoon Index. Because of the broader interest of such indices, particular care is taken to control the data quality, and so it is worth considering whether any of these datasets might be usable in public health contexts.

BOX 6.3 HOW DO WE MEASURE ENSO?

As discussed in Box 5.1, the Southern Oscillation is measured using an index of air pressure at Darwin and Tahiti. This Index is an indication of whether El Niño or La Niña are affecting the weather and climate. An advantage of the Index is that high-quality meteorological observations are available from 1866 and other observations can be used to estimate the Index back to 1829. However, a disadvantage of the Index is that it includes short-term weather variability, and so averaging over long periods (scientists use a five-month period) is required for a reliable indication of whether El Niño or La Niña is occurring.

El Niño and La Niña themselves are measured by averaging sea-surface temperatures over areas of the Equatorial Pacific Ocean and comparing these to long-term averages. The long-term average is updated, so that decades-long records are effectively detrended. Different areas have been defined for calculating the averages, and each region is labelled by a number; the most commonly used being the Niño3.4 region, which extends from close to the date-line towards South America (Figure 5.14).[19] Using measurements from ships, this index can be calculated back to 1949, and less reliably back to 1856. Since the 1980s, satellite measurements and data transmitted from a series of moored buoys have supplemented the ship observations to provide more accurate estimates. The Oceanic Niño Index (ONI) is the Niño3.4 index averaged over a three-month period, and is used operationally by some countries, including the USA, to define El Niño and La Niña events.

Numerous other indices exist,[20] but these are used more in research than in operational work. For most purposes, the Southern Oscillation Index (SOI), Niño3.4 or ONI are likely to be suitable. Similarly, and somewhat confusingly, different criteria are used to define whether El Niño or La Niña is happening, and so there is no universal agreement on when ENSO episodes have occurred, or agreement on when to declare a developing event. Nevertheless, the various classifications are similar.

6.3.2 Indirect measurements of climate by proxy, including by remote sensing

6.3.2.1 Historical proxy datasets

Climate scientists make extensive use of proxy measurements to infer changes in climate, primarily to reconstruct climate histories for times before the period of instrumental records. Examples include chemical and isotope analyses of ice cores, tree-ring data, coral growth and sedimentary deposits. Such datasets have been important for comparing recent global warming with previous warm epochs, for example. However, they are unlikely to be of direct interest for health analyses, and so are not discussed further here.

6.3.2.2 Satellite data

There are hundreds of satellites now in space, some of which have revolutionized developments in weather forecasting and climate work. Those of broad interest to meteorologists can be classified into two types according to their orbit, namely *polar-orbiting* and *geosynchronous* (also called *geostationary*). Polar-orbiting satellites, as the name implies, have an orbit which passes close to the North and South Poles. One advantage of such an orbit is that it is sun-synchronous – the satellite takes measurements for a given location at the same time of day (or 12 hours different) each overpass. Geostationary satellites, in contrast, remain vertically above a fixed point on the Earth's surface. Thus a geostationary satellite sees the same view of Earth all the time, and scans each point within that field of view every 15 minutes.

There are some important advantages of satellite data for climate work:

- *Cost*: some of the data are available free of charge.
- *Availability in real-time*: data from geostationary satellites are generally available in near real-time – often downloadable from the internet within 15 to 30 minutes.

- *Availability of historical data*: some datasets commence around the start of the satellite era, in the late 1970s, but high temporal and spatial resolution products are more recent.
- *Spatial coverage*: satellite imagery provides a spatially complete perspective for most of the world, including information for places with no *in situ* observations. This coverage is particularly important for countries whose environmental conditions, e.g., flooding, are determined by factors in neighbouring countries or where pests migrate over long distances.[21]

For many countries, satellite imagery provides the only affordable way of monitoring climate and environmental conditions in real-time. Satellite data are commonly used to map populations at risk of various environmentally sensitive diseases and to develop early warning systems.[22] However, because many of the environmental parameters are sensed indirectly, estimates may not be always reliable or usable in all circumstances and may require the interpretation of a skilled operator. The satellite estimates therefore need calibration and validation against ground-based data,[23] and so the use of satellite imagery does not negate the need for field measurements. As a result, many of the highest quality climate datasets are based upon a blend of satellite and station observations.[9]

Remote-sensing products are widely used by researchers studying infectious diseases that are influenced by climate and environmental factors. While a wide range of climate and environmental proxies are available for use in health studies[24] satellite data are most commonly used for estimating rainfall and temperature and for monitoring vegetation and water bodies. The accuracy and limitations of these remote-sensing products are discussed in turn.

6.3.2.2.1 Satellite monitoring of rainfall

No satellite yet exists that can reliably identify rainfall and accurately estimate the rainfall rate in all circumstances. Using a standard camera, a satellite can see the clouds from above that we see from below, but cloud presence by itself is not a good indicator of rainfall. Not all clouds produce rain, and rainfall intensity varies from place to place beneath those clouds that are generating rain. However, satellites carry sensors that can 'see' the Earth in a variety of ways (Box 6.4). Using a selection of sensors, in some situations it is possible to distinguish raining cloud from non-raining cloud by estimating:

- *Cloud-top temperatures*: deep convective clouds have cold, high tops, and so show up as low temperatures. This method of identification works best in the tropics and in the mid-latitude summer months when convective rainfall may predominate (§ 5.2.5.2). However, other types of rainfall (§ 4.2.2) may go unidentified because they do not form from cold clouds, and there may be false detection of rainfall from non-raining cold clouds (Box 6.4). Such errors may

BOX 6.4 REMOTE SENSORS

Sensors are instruments that are sensitive to different wavelength bands in the electromagnetic spectrum. The wavelengths in common use for rainfall, vegetation, soil properties, dust and land-surface temperature are summarized below.

Visible (wavelengths between 0.4 and 0.7 μm)

A sensor in the visible region of the spectrum acts like an ordinary camera and sees what your eyes would see if you were on the satellite. As visible radiation is just reflected sunlight, no information is obtained when the satellite is on the night-time side of the Earth. Visible radiation is used in cloud and vegetation monitoring.

- *Cloud monitoring*: thicker cloud is more opaque – i.e., less sunlight passes through it than through thin cloud and more is reflected back into space. Seen from below, the base of a thick cloud is dark, but seen from the satellite's point of view, the reflected light makes the cloud appear bright. Thinner clouds (whether at a low or a high altitude) allow through more light and reflect less. Therefore, thin clouds appear less bright to the visible sensor on the satellite.
- *Vegetation monitoring*: chlorophyll absorbs the red and blue electromagnetic wave and reflects the green. By measuring the reflectance at the sensor level in the red channel, it is possible to retrieve some information on the chlorophyll activity in the vegetation.

Near- and shortwave infrared (wavelengths between 0.7 and 3.0 μm)

Near-infrared wavelengths are sensitive to leaves while the shortwave infrared wavelengths are sensitive to water. Using combinations of the visible and infrared wavelengths it is possible to derive vegetation indices that provide specific information on the vegetation status. Some simple vegetation indices (such as the Normalized Difference Vegetation Index [NDVI]) have been developed to monitor vegetation greenness and quantity of biomass. A vegetation index is a simple mathematical formula used to estimate the presence of vegetation.

Thermal infrared (wavelengths between 8.0 and 15.0 μm)

The thermal infrared sensor registers radiation at wavelengths that are a good indicator of the temperature of the surface observed. The infrared image can be viewed during the day and during the night. The emission of infrared radiation is less at lower temperatures and greater at higher temperatures. The infrared sensor therefore acts as a remote thermometer which can estimate the temperature. However, their use in estimating air temperature is limited; instead, the temperatures are used to infer the presence of rain clouds. Because temperature decreases with height in a regular way depending on weather conditions and the amount of moisture present (§ 5.2.4), storm clouds show up as very cold with temperatures typically less than –40 °C and sometimes as low as –80 °C. Cirrus is also identified as very cold, and its presence is a problem when identifying rainfall using infrared sensors. Low-level coastal rain-producing and orographic clouds (§ 4.2.2) may show up at higher temperatures of between 0 °C and –20 °C, and so may also be problematic for rainfall estimation using infrared sensors.

Microwave (wavelengths between 1.0 and 300.0 mm)

Radiation emitted at microwave wavelengths is influenced strongly by the nature of the emitting surface (whether rough or smooth, wet or dry) and the size of particles through which it passes. Microwaves are scattered by water drops and ice crystals in cloud, and so, unlike the other forms of radiation already mentioned, microwaves can identify whether clouds have droplets big enough to produce rain. Thus, in general, microwave frequencies are better suited for rainfall detection than infrared frequencies. However, although rainy areas show up well over the oceans as bright against a dark background it is more complicated over the land because the background emission from the surface is very variable. The wetness of the surface as well as its roughness and the kind of vegetation all cause variations in the emitted radiation. Making quantitative estimates of rainfall against this continually changing background is a challenge. However, there are promising algorithms for overland rainfall estimation from microwave sensors that may provide a basis for the development of accurate rainfall datasets in the future.

be substantial in regions near the coast or in mountainous areas. Although estimates of rainfall from cloud-top temperatures have good spatial coverage, high temporal resolution and frequent updates (every 15–30 minutes), the accuracy is often poor

- *Cloud thickness*: the amount of water and ice in the cloud can be estimated by measuring the amount of scattered microwave radiation (Box 6.4). These methods offer a more accurate rainfall estimate, but have coarse spatial resolution and are updated only twice per day. Currently, the estimates are least accurate over the land, where, unfortunately, the information is needed most

CASE STUDY 6.2 SEASONAL MALARIA CHEMOPREVENTION IN THE AFRICAN SAHEL

Madeleine C. Thomson, IRI, Columbia University, New York, USA

The intense seasonality of the Sahelian rainy season, which runs from June–September, governs the lives and livelihoods of the region's diverse populations, many of whom are small-holder farmers and pastoralists (see Box 2.2). It also determines the short, but deadly, malaria season. These seasonal characteristics make the region an ideal choice for the use of presumptive seasonal malaria chemoprevention (SMC) with a combination of anti-malarial drugs as the key intervention.[25] When the transmission season is short (three to four months), anti-malarials administered monthly through the transmission season have the potential to protect children throughout the period of exposure to malaria infection – both at the individual level and through a population level 'mass effect'.

Compelling evidence from numerous studies as to the efficacy of SMC[26] prompted its rapid promotion to a regional programme for the Sahel. The target region for the SMC initiative was chosen on the basis that, on average, 60% of rainfall fell within three months.[27] Monthly Africa rainfall estimates (RFE 2.0) data for 2002–2009 at 10 km^2 spatial resolution were used to generate average long-term monthly rainfall, which were then used to define the average seasonality of each pixel in the region. Children under 5 who lived in areas that met the 60% criteria were then identified to be included in the implementation programme. The malaria season substantially follows the rains with peak malaria incidence lagging peak rainfall by about one to two months.[28]

Within a year, and with the support of the World Health Organization's (WHO) Technical Expert Group on Preventive Chemotherapy,[29] 3.2 million children under 5 years were protected from malaria by SMC in seven Sahelian countries – from Senegal to Chad.[27] Cost-effectiveness analysis for wide-scale implementation indicated that drug costs were marginal relative to distributional costs. Further studies revealed the significant benefits of the approach to children aged 6–10 years and that the extra budgets needed to include the additional children, while variable, were low relative to overall programme costs.[30] This was because older children resided in households where younger children were benefiting from the programme; substantially reducing the need for additional budgets for distribution.

In this example, freely available rainfall estimates for the entire Sahelian region are a pivotal aid in establishing where and when SMC should be implemented. Whether or not improvements in the quality of the rainfall data used would significantly improve programme efficacy is an open question.

Thus the images produced by the satellite are not much use for giving a precise estimate of rainfall for a particular spot on the ground at a particular time. The usefulness of satellites lies in their ability to give (literally) an overview, and the fact that they can be easy to access (see Case Study 6.2).

Techniques are being developed to take advantage of the better accuracy of microwave sensors and the better spatial and temporal coverage of infrared sensors by optimally combining the two products. Various monitoring products are becoming available using different ways of combining the sensor data, but it is not yet possible to generate a consistent historical set of rainfall estimates because some of the important sensors have been installed only on recent satellites.

Another approach towards generating better rainfall datasets is to blend the satellite rainfall estimates with available gauge measurements. The quality of the blended product depends on the quality, number, and distribution of the rain gauges used.[31,32] The ENACTS products (Case Study 6.1) are examples of blended data that are generated at the national level using a much larger set of station observations than are used in the generation of regional and global products. Currently, ENACTS products are available only for selected African countries.[33]

6.3.2.2.2 Satellite monitoring of temperature

The derivation of near-surface air temperature (typically measured at 2 m [§ 4.2.1]) from satellite sensors is far from straightforward. Clouds block the radiation from Earth's surface that the satellites measure to estimate temperature. Even in cloudless conditions when the satellites can measure the temperature of Earth's surface, the surface temperature is not necessarily a good indication of the air temperature (§ 5.2). Although night-time satellite products provide reasonable estimates of minimum temperatures, maximum temperature estimates are problematic.[34]

6.3.2.3 Data from drones

Unmanned aerial vehicles (UAVs), more commonly known as drones, are a new and significant technology with a wide range of potential uses at the environment-health interface. Their relatively low cost, small size and ease of use mean they can provide high resolution, rapidly updated information on very specific targets that service to fill a gap between ground-based stations and satellite-based sensors (see Box 6.5).

6.3.3 Modelled data

A key step in making weather forecasts is to use a weather prediction model to estimate the current conditions (§ 7.4.2). This estimate is known as an *analysis*, and represents our best estimate of the full state of the air in all three dimensions. Because of the motivation to be always improving the accuracy of weather forecasts, new and higher quality observations and upgraded models are frequently introduced in

BOX 6.5 DRONES

Michelle Stanton, Lancaster University, UK

Drones come in many shapes and sizes. Whilst they are perhaps more notoriously connected to military applications, their potential to benefit other industries and domains are increasingly being realized. For example, drones are being used as transportation devices, which, within the public health context, enable medical supplies to be delivered to remote health facilities and inaccessible disaster areas. More ambitiously, solar-powered drones are also being considered by companies such as Facebook as a method of delivering high-speed broadband internet to remote areas across the globe. Microsoft has been working on drones that are able to capture and discriminate between different mosquito species.

Within the context of climate and health, drones are starting to fill both the literal and figurative gap between climate and environment information obtained from ground-based stations and earth observation satellites. With the ability to fly close to the ground, access otherwise inaccessible areas, and carry multiple sensors, they bring access to real-time information on the climate and environment at the micro-scale at a much lower cost and greater accuracy than could be achieved using more conventional approaches. For example, cameras attached to a drone can collect contemporary images of the ground at a spatial resolution of a few centimetres, allowing vector control programmes to identify mosquito breeding sites, which can subsequently be targeted for larval source management.[35] Visual information collected by drones is also being used to explore the impact of a changing climate on biodiversity and agriculture, which also impact human health.

Whilst imagery of similar spatial resolution can be obtained from commercial satellites, its quality is often impeded by cloud cover and comes at a financial cost, making it a less attractive and affordable option for control programmes in resource poor settings. Publicly available data sources tend to be at a coarse spatial resolution and suffer the same limitations with respect to cloud cover, and subsequently temporal availability. The caveat to the use of drones for image capturing is that the image collection process is time consuming due to limitations in drone battery capacity and flight speed. However, as long as these images are being captured for local rather than large-scale operations the quality versus efficiency trade-off is often worth it.

Provided the carrying capacity of the drone is sufficient, it is, of course, not limited to capturing visual information. Sensors aboard drones are able to capture a wealth of climate information such as temperature, relative humidity, dust and even odours at altitudes and spatial scales of much greater relevance to human health than can be captured via satellites. This information can subsequently be integrated into early warning systems providing the ability to target health interventions spatially.

the process of producing the analysis (§ 6.1) The effect is that more recent analyses cannot be compared with older ones, and so it is problematic to use a history of analyses for climate work.

A few centres produce *reanalyses*, which are historical sets of analyses made using a fixed version of a weather prediction model, and a reasonably consistent set of observational data. Most reanalysis products have high temporal resolution (6-hourly), and reasonably high spatial resolution (< 2° latitude and longitude) and global coverage. There are some higher resolution regional reanalyses. Reanalyses provide excellent sources of information about the three-dimensional circulation of the air, and may be useful in studies such as the dispersion of dust and pollutants. For example, reanalysis data were used in a study of environmental drivers of meningococcal meningitis in the Sahel (Case Study 7.2).

Only a few surface observations are used in generating analyses and reanalyses, and so their estimates of the most commonly considered surface and near-surface parameters, such as rainfall and humidity, are problematic. Surface air temperature estimates from reanalyses are somewhat less problematic than rainfall because of the relatively large scale of temperature anomalies (§ 5.2.5), but should still be used with caution, and are no substitute for station-based observations.

Temporal changes in the observations are unavoidable when generating reanalyses, and so some consistency problems remain. As a result, reanalysis data are generally unsuitable for work on climate-change detection, for example. Some reanalysis products are not updated in real-time because the model may be outdated.

6.4 What data and information are available?

6.4.1 Availability of historical and real-time data

There is a bewildering range of climate datasets available. Selecting the most appropriate one(s) to use requires careful consideration of the required characteristics. Here we provide guidance on those characteristics; for information on specific datasets see reference 16. The most important selection criteria are likely to be:

- *Length of record*: station datasets provide the longest historical records; most satellite-based datasets start in the late-1970s or later.
- *Availability in real-time*: many satellite datasets are updated in near real-time; data from automated weather stations may be available in near real-time, but many station-based datasets are updated irregularly.
- *Spatial coverage*: satellite datasets provide complete global (but possibly omitting areas near the Poles), but represent area averages rather than location-specific values; interpolated station data can give the illusion of complete spatial coverage.
- *Spatial resolution*: satellite estimates of rainfall are available at high resolution, but are only accurate when averaged over large areas; the highest resolution satellite data are available for only a few years.

- *Temporal resolution*: monthly data are most widely available, but may mask extreme weather events.
- *Data quality*: station data provide ground-truthing, but may contain missing values and inhomogeneities; some satellite estimates may be difficult to verify, and may be of poor quality at high spatial and temporal resolutions.

Provision is made for open access to 'essential' meteorological data (Box 6.1), but, in practice, the accessibility and cost of historical data varies considerably from country to country. Fees are most often levied for daily and sub-daily data, and there can be long delays in accessing the latest observations. Monthly data are more widely available and can be useful in assessing seasonality, for example, but are of limited value for risk assessment of hazardous events, with the exception of slow-onset hazards such as drought.

In general, the highest quality and most abundant datasets are for rainfall (especially in the tropics) and temperature (especially in the extratropics). When using rainfall data for public health research, care needs to be taken to avoid assuming that rainfall is a good indicator of flooding. Floods have a number of typologies, including coastal, riverine and flash floods (Box 6.6). Coastal floods may arise from sea-level rise, and riverine and flash floods may result from snowmelt or from heavy rainfall. In the case of snowmelt, the snow may have fallen far from the flooded

BOX 6.6 FLOODING

Andrew Kruczkiewicz, IRI, Columbia University, New York, USA

A flood includes the notion of water being present over a land surface where and when it usually does not occur. The specific definitions of flood events are complex and depend on the source and path of the water. It is important to understand the different characteristics of flood types (see Table 2.1) because the methods of forecasting both the hazard and impact will differ.

One of the most common types of floods, *riverine flood*, results in relatively large areas of standing water over land that is usually not covered with water. For example, the Sudd is a vast swamp in South Sudan, formed by the White Nile's *Baḥr al-Jabal* section. High water levels in Lake Victoria, the source of the While Nile, in the early 1960s resulted in a tripling of the size of the swamp during 1961–1963 and the loss of extensive tracks of *Acacia-Balanities* woodland. The regrowth of the forests in the 1980s was identified as a contributing factor to the epidemics of visceral leishmaniasis that devastated the region.[36]

Most incidents of riverine flooding occur for much shorter periods of time – from days to weeks and their health impacts are more immediate. While flooding may occur as a result of extreme or persistent rainfall at the same location the source of the floodwaters may come from rainfall or snowmelt occurring hundreds of miles away. Floodwaters can lead to increased risk for water-borne diseases. Heightened risk of impacts can last for days to weeks after the initial 'flood' or inundation occurs.

Flash floods are characterized by rapid movement of water in a relatively local area leading to risk of impact from building collapse, bridges destroyed, debris flow and rapid onset disruption of sanitation systems. Drowning and crush injuries are common with flash floods.

Coastal flooding is caused by the intrusion of sea water over land and can occur without any rainfall. For example, storm surge (the rise and push of sea water over land) usually occurs to some extent during tropical cyclones, however the location of storm surge, especially the area of maximum storm surge, can occur at locations that are distant from the landfall point and can occur in areas that receive little rainfall. As a multi-hazard event, tropical cyclones can cause different types of flooding in different areas and at different times. For example, during Hurricane Harvey in the 2017 North Atlantic hurricane season, heavy rainfall caused the most severe flash flooding in inland areas in Houston. However, the landfall of the storm occurred 240 km south of Houston and led to storm surge in that region near Rockport, TX. Understanding the timing and distribution of different types of floods is important for the health sector to better pre-position medical resources before potential impact, particularly in a multi-hazard event with different types of floods.

areas weeks or even months before the flooding occurs. Some floods are entirely predictable, as when water is released from an upstream dam.

Most other data of likely interest in public health work (e.g., wind speeds and humidity) are harder to come by than for rainfall and temperature, and data quality may be problematic.

Regardless of the ease of data access, the selection, use and interpretation of climate data requires considerable expertise. Any application of climate data for public health is best accomplished in partnership between the public health and climate communities.

6.4.2 Availability of historical and real-time information

Accessing climate data may be important for many research purposes, but access to information products in practical decision-making may be more useful than access to the underlying climate data per se. As with the climate datasets, there is a large and growing selection of weather and climate monitoring products available. Most of these products are targeted at meteorologists, but only a small subset are presented in (relatively) easy-to-understand formats with broader audiences in mind, and fewer still are tailored specifically for public health specialists.

Monitoring of hazardous weather conditions is a forecasting as well as a monitoring problem, and so is discussed in Chapter 7. However, for slow on-set disasters and persistent anomalies such as drought, climate monitoring products can be useful even when it is not possible to provide accurate weather or climate forecasts. In addition, in-built lags in the transmission dynamics of many infectious diseases means that forecasts of disease incidence can be created from monitoring cumulative climate and environmental conditions (see § 3.4).

The international coordination of a climate monitoring infrastructure has received less attention than for forecasting, but addressing this oversight is a high priority of the GFCS (see Box 1.3). For example, the WMO is actively encouraging the implementation of climate watches, similar to weather alerts (§ 7.6.1), to provide official notifications of severe climate conditions.[37] Climate watches are currently implemented in only a few countries, although routine information products are more widespread. For example, ten-day and monthly bulletins, and annual reports are produced by many countries. In many cases, these bulletins are targeted at the agricultural sector, or are written for an expert meteorological readership, and few are likely to be accessible to a public health audience. The situation is beginning to improve, most notably for drought and heat (§§ 6.4.2.1 and 6.4.2.2), and a few climate monitoring products are starting to be developed specifically for health specialists. Some examples are available from the International Research Institute for Climate and Society (IRI), and from a few NMHSs. Meanwhile, the more generic bulletins may still fulfil important monitoring functions by providing information on extreme events that are ongoing or that have occurred recently and whose full impacts may yet to be experienced.

Annual climate reports are intended to serve a more retrospective purpose than the bulletins. The reports may be useful for gauging the impacts of any major climate events over the previous year, or for obtaining a sense of the severity and expected frequency of such events. The availability and usefulness to health specialists of the bulletins and reports will vary considerably from country to country depending largely on national capacity. If national level information is unavailable or inadequate, the Regional Climate Centres (§ 6.4) should be able to provide some information, since climate monitoring is one of their mandatory functions. A list of these centres is available from the WMO.[iv] Temperature and rainfall are monitored by all the centres, but additional information on extremes and climate impacts such as flooding may be available.

An important function of the annual and monthly reports is to provide regular updates to the less frequent IPCC Assessment Reports (see § 9.2) that review, inter alia, evidence for how climate is changing. Although climate-change monitoring is the dominant theme of the National Oceanic and Atmospheric Administration's (NOAA) monthly Global Climate Reports, these Reports also provide updates on major climate phenomena such as El Niño (Box 5.1). Occasional Special Reports are released after weather and climate events of particular note. All these various reports are key inputs to the WMO's Annual Statement on the Status of the Global Climate that is intended for a broad audience.

There are a few monitoring products that combine meteorological data with other environmental data to provide information that is intended to relate closely to impacts. Examples include products that target drought and air quality.

6.4.2.1 Drought monitoring

There are multiple types of drought, depending on whether the focus is on the rainfall deficit, the water deficit or the impacts of either (or both) deficits.[38] Further, there are multiple ways of measuring each particular type of drought (see §§ 2.2 and 4.3.3 on measures of rainfall deficit). Regardless of the broad array of definitions, national drought information products are available in many countries, and there are many regional and global drought monitoring products. Most of these products are based on rainfall deficits, but some include soil moisture, impacts on crops and other vegetation, fire risk, and water supply. Drought monitoring that focuses on food insecurity may be of particular interest to health specialists because of the effect of poor nutrition on health outcomes. For example, the Famine Early Warning Systems Network (FEWSNET)[39] combines data on rainfall with a range of socioeconomic data on vulnerability to map food insecurity in parts of Central America and the Caribbean, much of Africa, and part of Central Asia.

6.4.2.2 Air chemistry and air quality monitoring

The monitoring of greenhouse gases is primarily a climate-change question, and is discussed in § 9.1. Urban air pollution monitoring and information dissemination is conducted primarily at the national scale, but some regionally coordinated monitoring initiatives have been established because of transboundary pollution issues. Examples include regional smoke haze resulting from land and forest fires (monitored by the Association of Southeast Asian Nations [ASEAN] Specialized Meteorological Centre), and mineral dust in West Africa (monitored by the Sand and Dust Storm Warning Advisory and Assessment System of the Northern Africa-Middle East-Europe Regional Centre). The mineral dust in West Africa is of natural origin, but is of interest because of its links to epidemics of bacterial meningitis (see Case Study 7.1).

6.5 Conclusions

There is a bewildering array of climate datasets available, but rarely does one dataset stand out as clearly superior to all the others. To match data to decision-maker needs careful choices have to be made between duration, real-time accessibility, spatial coverage, resolution and quality. Satellite and other remotely-sensed observations have dramatically expanded the options, but do not negate the need for station observations because satellite estimates require calibration against ground-based data. Regardless of which dataset is used, ultimately, information about the past and present is used to make inferences about the future weather and climate. In many cases, predictions may be available; these are discussed in the following chapters, beginning with forecasts for the next few days.

Notes

i Climate Hazards group InfraRed Precipitation with Satellite data.
ii Tropical Applications of Meteorology using SATellite and ground based observations.
iii http://public.wmo.int/en/our-mandate/what-we-do/data-exchange-and-technology-transfer.
iv www.wmo.int/pages/prog/wcp/wcasp/rcc/rcc.php.

References

1 Thomson, M. C. *et al.* Using rainfall and temperature data in the evaluation of national malaria control programs in Africa. *American Journal of Tropical Medicine and Hygeine* **97**, 32–45, doi:10.4269/ajtmh.16-0696 (2017).
2 Manton, M. J. *et al.* Observation needs for climate services and research. *Procedia Environmental Sciences* **1**, 184–191 (2010).
3 Branski, F. Pioneering the collection and exchange of meteorological data. *World Meteorological Bulletin* **59**, 12–17 (2010).
4 Trewin, B. The status of the Global Observing System for climate. *World Meteorological Bulletin* **65**, 48–53 (2016).
5 Peterson, T. C. & Rose, R. S. An overview of the global historical climatology network database. *Bulletin of the American Meteorological Society* **78**, 2837–2849 (1997).
6 Trenberth, K. E. *et al.* in *Climate Science for Serving Society*, 13–50 (Springer, Dordrecht, 2013).
7 IPCC. *Climate Change* 2014*: Synthesis Report. Contribution of Working Groups I, II and III to the Fifth Assessment Report of the Intergovernmental Panel on Climate Change* [Core Writing Team]. 151pp (IPCC, Geneva, 2014).
8 Hewitt, C., Mason, S. & Walland, D. The global framework for climate services. *Nature Climate Change* **2**, 831–832 (2012).
9 Dinku, T. *et al. THE ENACTS APPROACH: Transforming Climate Services in Africa One Country at a Time* (World Policy Institute, New York, 2016).
10 Kobayashi, S. *et al.* The JRA-55 reanalysis: general specifications and basic characteristics. *Journal of the Meteorological Society of Japan* **93**, 5–48, doi:10.2151/jmsj.2015-001 (2015).
11 WMO. *World Meteorological Organization, Convention, General Regulations, Staff Regulations, Financial Regulations and Agreements 15* (WMO, Geneva, 2015).
12 Zillman, J. W. Atmospheric science and public policy. *Science* **276**, 1084–1086 (1997).
13 WMO. *Meteorological Data: Guidelines on Relationships in Commercial Meteorological Activities* (WMO, Geneva, 1996).
14 WMO. *WMO Global Atmosphere Watch (GAW) Implementation Plan:* 2016–*2023* (WMO, Geneva, 2017).
15 Schultz, M. G. *et al.* The Global Atmosphere Watch reactive gases measurement network. *Elementa: Science of the Anthropocene 3*, 000067 (2015).
16 Mason, S., Kruczkiewicz, A., Ceccato, P., & Crawford, A. *Accessing and Using Climate Data and Information in Fragile, Data-poor States* (International Institute for Sustainable Development, Winnipeg, MB, Canada, 2015).
17 Cooper, J. Rescue, archive and stewardship of weather records and data. WMO Bulletin. *Bulletin of the World Meteorological Organisation* **64**, 28–31 (2015).
18 Omumbo, J., Lyon, B., Waweru, S. M., Connor, S. & Thomson, M. C. Raised temperatures over the Kericho tea estates: revisiting the climate in the East African highlands malaria debate. *Malaria Journal* **10**, 12, doi:10.1186/1475-2875-10-12 (2011).

19 Trenberth, K. E. The definition of El Niño. *Bulletin of American Meteorological Society* **78**, 2771–2777, doi:10.1175/1520-0477(1997)0782.0.CO;2 (1997).

20 Trenberth, K. E. & Stepaniak, D. P. Indices of El Niño evolution. *Journal of Climate* **14**, 1697–1701, doi:10.1175/1520-0442(2001)0142.0.CO;2 (2001).

21 Ceccato, P., Cressman, K., Giannini, A. & Trzaska, S. The desert locust upsurge in West Africa (2003–2005): information on the desert locust early warning system and the prospects for seasonal climate forecasting. *International Journal of Pest Management* **53**, 7–13, doi:10.1080/09670870600968826 (2007).

22 Ceccato, P. *et al.* Integrating remotely-sensed climate and environmental information into public health in *Integrating Scale in Remote Sensing and GIS* (eds. Quattrochi, D. A., Wentz, E., Lam, N S-N. & Emerson, C. W.), 304–335 (CRC Press, Florida, 2016).

23 Dinku, T., Connor, S. J., Ceccato, P. & Ropelewski, C. F. Intercomparison of global gridded rainfall products over complex terrain in Africa. *International Journal of Climatology* **28**, 1627–1638 (2008).

24 Ceccato, P. *GEO Task US-09-01a: Critical Earth Observation Priorities, Human Health: Infectious Diseases Societal Benefit Area.* 149pp (GEO, Geneva, Switzerland, 2010).

25 Cairns, M. *et al.* Estimating the potential public health impact of seasonal malaria chemoprevention in African children. *Nature Communications* **3**, 881, doi:10.1038/ncomms1879 (2012).

26 Meremikwu, M. M., Donegan, S., Sinclair, D., Esu, E. & Oringanje, C. Intermittent preventive treatment for malaria in children living in areas with seasonal transmission. *The Cochrane Library* **2**, CD003756, doi:10.1002/14651858.CD003756.pub4 (2012).

27 Noor, A. M. *et al.* Sub-national targeting of seasonal malaria chemoprevention in the Sahelian countries of the Nouakchott Initiative. *PLOS One* **10**, e0136919, doi:10.1371/journal.pone.0136919 (2015).

28 Jusot, J.-F. & Alto, O. Short term effect of rainfall on suspected malaria episodes at Magaria, Niger: a time series study. *Transactions of the Royal Society of Tropical Medicine and Hygiene* **105**, 637–643 (2011).

29 WHO/GMP. *Report of the Technical Consultation on Seasonal Malaria Chemoprevention (SMC)* (WHO/GMP, Geneva, 2011).

30 Pitt, C. *et al.* Large-scale delivery of seasonal malaria chemoprevention to children under 10 in Senegal: an economic analysis *Health Policy and Planning* **32**, 1256–1266 (2017).

31 Dinku, T., Ceccato, P., Grover-Kopec, E., Lemma, M., Connor, S. J. & Ropelewski, C. F. Validation of satellite rainfall products over East Africa's complex topography. *International Journal of Remote Sensing* **28**, 1503–1526 (2007).

32 Dinku, T., Chidzambwa, S., Ceccato, P., Connor, S. J. & Ropelewski, C. F. Validation of high-resolution satellite rainfall products over complex terrain. *International Journal of Remote Sensing* **29**, 4097–4110 (2008).

33 Dinku, T., Cousin, R., del Corral, J., Ceccato, P., Thomson, M., Faniriantsoa, R. *et al.* The ENACTS Approach: Transforming climate services in Africa one country at a time. New York: World Policy Institute, 2016 March 2016. https://worldpolicy.org/wp-content/uploads/2016/03/The-ENACTS-Approach-Transforming-Climate-Services-in-Africa-One-Country-at-a-Time.pdf

34 Vancutsem, C., Ceccato, P., Dinku, T. & Connor, S. J. Evaluation of MODIS Land surface temperature data to estimate air temperature in different ecosystems over Africa. *Remote Sensing of Environment* **114**, 449–465 (2010).

35 Hardy, A., Makame, M., Cross, D., Majambere, S. & Msellem, M. Using low-cost drones to map malaria vector habitats. *Paasite and Vectors* **10**, 29, doi:10.1186/s13071-017-1973-3 (2017).

36 Ashford, R. W. & Thomson, M. C. Visceral leishmaniasis in Sudan – a delayed development disaster. *Annals of Tropical Medicine and Parasitology* **85**, 571–572 (1991).

37 WMO. *Climate Watch System Early Warning Against Climate Anomalies and Extremes* (WMO, Geneva, 2006).

38 Heim Jr, R. R. A review of twentieth-century drought indices used in the United States. *Bulletin of the American Meteorological Society* **83**, 1149–1165 (2002).

39 Verdin, J., Funk, C., Senay, G. & Choularton, R. Climate science and famine early warning. *Philosophical Transactions of the Royal Society B-Biological Sciences* **360**, 2155–2168 (2005).

7

WEATHER FORECASTS

Up to one week in advance

Simon J. Mason and Madeleine C. Thomson
Contributors: Heat Action Group, Kim Knowlton, Hannah Nissan, Ángel G. Muñoz, Carlos Perez Garcia-Pando and Jeffrey Shaman

> Home without boots, and in foul weather too!
> How scapes he agues, in the devil's name?
>
> Henry IV Part I *by William Shakespeare*

7.1 Introduction

A waiter at a dinner party once asked the Greek philosopher Socrates:

> 'Are you, Socrates, the one people consider the Expert?'
> 'Well, is that not better,' Socrates replied, 'than being considered the Idiot?'
> 'It would be were you not considered an expert meteorologist.'
>
> (Xenophon, *Symposium* 6)

Of course, in Socrates's time, a 'meteorologist' was not quite the physics nerd that we may imagine today, but regardless of what 'meteorologists' are expert at, they have had bad reputations for centuries. Whereas Socrates was criticized for encouraging speculative and subversive thinking, today meteorologists are criticized from a sense that their weather forecasts are not as accurate as they should be.

Notwithstanding these negative public perceptions, weather forecasters have long been partners with public health practitioners involved in disaster management associated with extreme events. However, it is only relatively recently that weather forecasts have been considered as a public health tool. In recent years, National Meteorological and Hydrological Services (NMHSs) have helped develop tailored health forecasts designed for routine use by the public, hospital managers and general practitioners (GPs).[1] These forecasts provide early warning of increases in illnesses that may be linked to changes in the weather – such as heart attacks, strokes, respiratory diseases, infectious diseases and broken bones. For example, the UK Met Office, in collaboration with the UK's National Health Service, has developed a weather-based alert system for Chronic Obstructive Pulmonary Disease (COPD), which has been shown to reduce hospital visits.[2] Such tailored forecasts can only be created in partnership with health specialists, meteorological agencies and health authorities. As another example, the Shanghai Meteorological Bureau has established co-operative agreements with many partners including: the Shanghai Food and Drug Supervision Administration for joint release of warnings of bacterial food poisoning; and the Shanghai Health Bureau and the Shanghai Municipal Center for Disease Control and Prevention for joint dissemination of heat wave information and human health monitoring and warning systems.[3] Such partnerships are becoming more common, indicating a significant recent shift in the long history of public distrust of weather forecasters.

In this chapter, we examine how weather forecasts (see Box 7.1) are made and how they may be used by the health sector. We explore why, where and when we might obtain good forecasts and why sometimes the forecasts go wrong. We consider the theoretical and practical constraints on weather forecasts and why their accuracy declines rapidly after only a few days. Knowing the limitations of weather forecasts helps us to learn how to make best use of such information.

7.2 Why weather forecasts may be useful to the health community

Weather forecasts are particularly valuable in predicting *hazardous* weather conditions that put lives at risk, including the hydro-meteorological disasters outlined in Table 2.1 in Chapter 2. Forecasts of heavy rainfall may provide early warning of floods and landslides enabling early evacuation of at-risk households. They may also indicate relief to firefighters combatting wildfires. Storm warnings give time for individuals to seek shelter and authorities to prepare relief. For example, fishermen on Lake Victoria have long been vulnerable to localized storms; weather forecasts are now playing an increasing role in keeping them safe.[4] Dust-storm warnings indicate reductions in visibility and air quality and encourage people (especially children) to stay indoors to avoid respiratory problems such as asthma or bacterial meningitis (see Case Study 7.1). Warnings of cyclones (including hurricanes, typhoons, and winter storms; § 4.2.8), which can have devastating effects over large areas, may result in the mass evacuation of populations or the shutdown of significant economic activity.

Forecasts of extreme and persistent temperatures (both heat waves and cold snaps) provide warnings of *inhospitable* weather conditions for which local populations are not adapted and where, without some form of protection, human survival will be compromised. Following the 2003 heat wave in Europe, which resulted in the deaths of over 70,000 people,[10] Heat Early Warning Systems (HEWS) were established in many countries.[11] To be effective a HEWS must be based on locally determined critical meteorological thresholds, above which the adverse impacts of heat begin to escalate. Such thresholds vary from region to region according to

BOX 7.1 WEATHER AND CLIMATE FORECASTS

A *weather forecast* can be a statement of:

- What the atmosphere will be like at a specific time in the future (although the exact time of day may not be specified); for example, at the time of writing, the temperature tomorrow is predicted to drop to –1 °C in New York, and snow is expected (with a probability of 80%).
- A specific atmospheric state at some uncertain time in the future; for example, at the time of writing (on a Thursday) a winter storm is expected to affect New York sometime around Monday or Tuesday next week.

A *climate forecast* is a statement about the general state of the atmosphere over a prolonged period, but makes no mention of what the weather will be like at any particular time during that period.

Weather and climate forecasts consist of:

- *A predictand*: the specific meteorological parameter (§ 4.2) that is being predicted, and that must be measured or estimated to test how accurate the forecast was. For a weather forecast, the meteorological parameter is often expressed as a maximum or minimum (in the case of temperature) or a total (in the case of precipitation); for a seasonal forecast, some aggregate (e.g., an average, an accumulation, a count) of the meteorological parameter is used. Sometimes the predictand is a specific weather phenomenon, usually hazardous, such as a tornado or storm, which must be described using multiple meteorological parameters.
- *A time- or target-period*: the period for which the forecast applies. A weather forecast typically refers to a specific date or time of day, whereas a seasonal forecast refers to a specific year and a three- or four-month period.
- *A lead-time*: the gap in time between the date the forecast is made and the start date of the period being forecast. The lead-time is how long you have to wait before the time the forecast is targeting begins. For example, if it is noon now, a weather forecast for tomorrow afternoon has a lead-time of 24-hours; if it is early March, a seasonal forecast for April–June will have a lead-time of about one month. Together, the lead-time and the time-period determine the *range* of the forecast. The range indicates how far into the future the forecast extends. The World Meteorological Organization (WMO) has set definitions of various forecast ranges (Table 7.1). According to these definitions, seasonal forecasts (Chapter 8) are examples of long-range forecasts, whereas sub-seasonal forecasts (Box 7.5) are examples of extended-range weather forecasts.
- *An indication of the uncertainty*: a forecast is most useful if it involves: a) a statement of what is expected to happen, as well as b) an indication of

how confident the forecaster is. There are different ways of communicating this degree of confidence (see Box 7.4).

TABLE 7.1 Definitions of meteorological forecasting ranges

Range	*Definition*
Nowcast	0–2 hours
Very short-range weather forecast	< 12 hours
Short-range weather forecast	12–72 hours
Medium-range weather forecast	72–240 hours
Extended-range weather forecast	11–30 days
Long-range forecast	30 days–2 years

Source: WMO[i]

CASE STUDY 7.1 DUST STORM IMPACTS ON HEALTH

Carlos Perez Garcia-Pando, Barcelona Supercomputing Center, Barcelona, Spain

Dust storms are meteorological hazards that arise when strong winds blow loose sand and dust from a dry surface. They affect many arid and semi-arid regions of the world including North Africa and the Sahel, southern Europe, the Middle East, central and East Asia, Australia and the western United States. The airborne dust emitted from these regions is a key atmospheric constituent and represents an important natural source of atmospheric particulate matter. Dust impacts the climate, the weather, ecosystems and air quality (and consequently economic activities and human health). The Sand and Dust Storm initiative of the World Meteorological Organization[5] brings together global researchers and operational meteorologists whose models of dust emission, transport and deposition are used to predict dust storms days before they occur; providing authorities with several days' lead-time to take mitigating actions.

The region in the world most impacted by dust storms is the African Sahel. Much of the dust that forms the dry season storms originates in the ephemeral lake beds of the Bodélé Depression in Chad. Transported on the north-easterly Harmattan trade-wind (that blows from the Sahara into the Gulf of Guinea) these fine mineral particles are a risk factor for a wide range of health issues including cardiovascular disease and respiratory problems.

Dust has long been suspected as a risk factor for epidemic meningococcal meningitis, an infectious bacterial disease that is particularly common in the Sahel during the dry season. In 2007 the Meningitis Environmental Risk Information Technologies (MERIT) initiative, led by the World Health Organization, was established to identify opportunities for health and climate/dust scientists to work together to develop a meningitis epidemic early warning system for the Sahel.[6] The prediction of epidemics is challenged by the lack of spatially and temporally resolved near-real time information on the levels of carriage of the bacteria, population immunity and serogroup type and virulence. Although weather/climate and other environmental data contain uncertainties, they have shown the potential to improve meningitis forecasting. Several studies have established that statistical models including weather and dust aerosol data as inputs, could potentially forecast the risk of epidemics, saving lives and effectively allocating scarce vaccine resources.[6–8] A study in Niger[9] used a combination of epidemiological data and mechanistic studies in mouse infection models to investigate the link between climate and invasive bacterial disease. Disease surveillance and climate monitoring revealed that high temperatures and low visibility (a proxy for dusty air) were significant risk factors for bacterial meningitis and that the bacteria *Streptococcus pneumoniae* was more invasive in mice exposed to dust or heat. The dust effects were associated with a reduced ability of white blood cells to direct bacterial killing while high temperatures increased the release of damaging bacterial toxins.

the climate and the vulnerability of the local population and can be determined through research using local climate and health data,[12,13] or through surveys and community consultations.[14] Weather forecasts can then be tailored to indicate the likelihood of reaching these thresholds, and should be issued at lead-times appropriate to the local capacity to take action, within the limits imposed by forecast skill (see Box 7.2 and 7.3 and Case Study 7.2). A set of standard operating procedures outline the measures to be taken in response to the forecasts. These procedures vary depending on vulnerable population groups, the local capacity of stakeholders to respond, the available resources and the skill of the forecast.

BOX 7.2 SUB-SEASONAL FORECASTS

Ángel G. Muñoz, IRI, Columbia University, New York, USA

A timescale of great interest is the sub-seasonal one, typically involving forecasts of weekly averages of variables like rainfall or temperature, a few weeks ahead. Sub-seasonal forecasts are particularly relevant to the prediction of heat waves – providing a potential bridge between seasonal risk assessments, seasonal forecasts and weather warnings (see Box 7.3).

Lead-times of sub-seasonal forecasts are long enough that much of the information in the initial conditions is lost, but at the same time are too short for other sources of predictability (such as sea-surface temperatures) to have a strong influence. Presently, sub-seasonal skill is limited, and in general sub-seasonal forecasts cannot yet be used to develop climate services. There are different physical processes that are sources of predictability at sub-seasonal-to-seasonal timescales. These processes include: interactions between the tropics and the extra-tropics that are related to tropical heating;

persistent weather patterns, like blocking; persistent oceanic conditions in the extra-tropics and the tropics; persistence of soil moisture anomalies affecting the overlying air; large-scale weather patterns like the Madden-Julian Oscillation (MJO, see Box 5.2); interactions between different weather and climate phenomena acting at multiple timescales, e.g., MJO and the El Niño – Southern Oscillation (ENSO, see Box 5.1), can enhance skill compared to when they are acting independently.[15]

BOX 7.3 POTENTIAL USE OF SUB-SEASONAL FORECASTS IN HEAT EARLY WARNING SYSTEMS

Hannah Nissan, IRI, Columbia University, New York, USA

Skilful forecasts are an essential building block of a HEWS, but for these forecasts to be actionable they must be tailored to the needs of decision-makers, focusing on critical thresholds, provided on appropriate lead-times, and calibrated for reliability so that forecast confidence can be easily interpreted.

Many risk reduction actions require more advanced warning than the few days provided by weather forecasts. Table 7.2 illustrates some of the measures that could be taken if heat wave forecasts could be made available on longer lead-times. In some ways, heat waves present a good test case for the use of sub-seasonal forecasts, which are currently only experimental, in early warning systems. Adaptation measures for hydro-meteorological hazards like flash floods tend to involve costly actions like evacuations and flood defence reinforcements. In contrast, there are many low-cost, no-regret adaptation options appropriate to cope with extreme heat, from training for health staff and social workers, to procuring emergency drinking water and contingency planning for outdoor sporting events. Furthermore, predictability of heat waves has been demonstrated on sub-seasonal timescales for several regions including Europe, South Asia and the United States.[12,16,17]

At longer lead-times, reliable probabilistic forecasts (Box 7.4) become even more critical for an effective early warning system. Reliable forecasts are essential because they allow decision-makers to pair appropriate actions with forecast skill. Low-regret interventions such as recapping emergency response procedures or closely monitoring weather forecasts can be paired with weak probability forecasts. Actions that incur a higher cost, which might include rescheduling outdoor sporting events or setting up emergency drinking water stations, can be contingent on a higher probability trigger.

TABLE 7.2 Actions that could be taken in response to heat wave warnings at different timescales (sub-seasonal is in italics as it indicates potential only)

Ready	*Set*	*Go*
Seasonal risk Actions to take every year before the at-risk season	Higher risk in next month Actions to be taken following sub-seasonal climate forecasts (if and when available)	Imminent risk Actions to be taken following a weather alert
Review average heat risk across season	*Monitor weather forecasts closely*	Prepare utilities for increased power demand
Access seasonal forecast	*Re-cap emergency action plan*	Prepare to open cooling centres
Update heat action plan	*Inform schools*	Begin mass media and public awareness campaign
Review communication strategies	*Inform cooling centres*	Distribute emergency drinking water
Refresh medical and media training	*Reinforce co-ordination with disaster management personnel*	Alert hospitals to increased demand
Prepare contingency plans for events	*Distribute appropriate advice through media*	Reschedule hospital staff shifts
Supply routes for backup water and generators	*Procure emergency drinking water*	Check in on elderly
Coordinate with utilities to ensure continued power supply		Reschedule sporting events

CASE STUDY 7.2 HEAT ACTION PLANS AND EARLY WARNING SYSTEMS HELP SAVE LIVES IN INDIA

Kim Knowlton, Mailman School of Public Health, Columbia University, New York, USA

India currently faces a challenging array of climate-health threats, and among the most prominent are heat waves. Heat waves in 2010 and 2015 killed thousands of people in India. The Public Health Foundation of India (PHFI) and the Indian Institute of Public Health-Gandhinagar (IIPH-G), in partnership with the Natural Resources Defense Council (NRDC), and under the leadership of the Ahmedabad Municipal Corporation (AMC), have successfully developed a set of strategies to counter the health challenge of extreme heat. The first municipal Heat Action Plan (HAP) and heat early warning system in South Asia was launched by AMC with the help of NRDC, IIPH-G and partners in the western Indian city of Ahmedabad, in Gujarat state, in 2013.[14]

The heat early warning system initially produced probabilistic forecasts of maximum temperature, together with probabilities that the temperature would reach specific mortality-determined temperature thresholds. The thresholds applied in the pilot plan were based on group consensus after analysis of the 2010 heat wave mortality and temperature data. Initially, probabilistic forecasts had a seven-day advance lead-time, which was identified during initial planning discussions with AMC as most useful to provide early warning and mount a coordinated public health and inter-agency response.

The daily temperature forecasts were sent via e-mail alerts to the AMC's 'nodal officer' who served as point person on inter-agency notification and coordination. Besides daily maximum temperatures, a second panel on the forecasts provided a summary for the previous seven days, which was intended for decision-makers' use to show forecast consistency and highlight recent heat threat levels.

After a successful 2013 pilot phase, by 2016 another ten additional cities had adopted the Ahmedabad HAP developed by NRDC and partners to prepare for extreme heat, raising the numbers of people served by these HAPs from seven million to 15 million. By mid-2017, 17 cities and 11 states had adopted or were developing HAPs in India. At the national level, the Indian Meteorological Department (IMD) and the Indian Meteorological Society (IMS) have stepped up with five-day maximum temperature deterministic forecasts expanded to over 300 cities, enabling many more cities to build local HAP pilot projects, even without the probabilistic forecast features. The National Disaster Management Authority (NDMA) has issued new national guidelines and television advertisements in local languages focused on protecting communities from extreme heat.

The municipal framework for protecting communities from the health effects of extreme heat is also being adapted to the State level. Key components include developing an early warning system, building capacity among public health professionals in the state, building community and public awareness, and effective interagency coordination with support of the disaster management authorities. By helping partners in India to implement state-level HAPs, NRDC's goal is to help protect the health of hundreds of millions from heat-health threats. This work is enhancing climate resilience in India, reducing heat vulnerability today and for the future, and helping to save lives.

Forecasts of *unhealthy* weather (poor air quality) are now common in many cities around the world. Under certain circumstances, meteorological conditions promote ozone production and, combined with atmospheric pollutants and dust, create a toxic environment for plant, human and animal respiration. Forecasts of air quality are increasingly important in many urban environments.[18]

Finally, forecasts may be for weather conditions that are *suitable* for supporting a wide range of pathogens, pests and diseases that are important for crop production and/or human and animal health. When considering the development of a weather-driven health forecast system, it is important to consider exactly what weather parameter needs to be forecasted. It is also important to know exactly how good forecasts are (Box 7.4) because weather forecasts may result in expensive and time consuming responses for individuals, communities, government agencies and private businesses.

Accurate weather forecasts can provide a timely warning of a weather event with potentially severe health consequences. The lead-time for action provided by the forecast is critical to its potential value. Tornado warnings may provide minutes of valuable time for individuals to shelter in their basement, while hurricane warnings provide several days for hospitals to move critical care patients away from the storm's likely path.

With a good climate monitoring system in place, confident early warnings of lagged health impacts (such as vector-borne disease epidemics) may be possible (see § 3.4). The added lead-time provided by weather forecasts in predicting the meteorological triggers of lagged health events may not be worthwhile. They often do not add much lead-time to the monitoring information and, of course, they are less certain than observing the weather event itself. Weather forecasts may, however, still be useful in predicting potential problems in the supply chain that is needed to respond to lagged events. For example, weather forecasts may be useful for anticipating possible damage to infrastructure (e.g., washed-away bridges) from rain-induced flooding, which may result in disrupted supply chains for commodities (drugs, drips, protective gloves, etc.) needed to counter an epidemic. Knowing the accuracy or reliability of the forecasts is a prerequisite to identifying the best ways of managing the risk (the distinction between accurate and reliable forecasts is explained in Box 7.4).

BOX 7.4 MEASURING HOW GOOD (OR BAD) FORECASTS ARE

The measurement of how well forecasts compare with the observed outcomes is called *forecast verification.*[19] Measuring how good a forecast is (or a set of forecasts are) is a field of scientific research in its own right.[20] Here we provide only the briefest introduction to some of the most important terms and concepts.

How forecasts are verified depends on how the outcome is measured (is the outcome discrete, as with rainfall occurrence, or continuous, as with temperatures?) and on how the forecasts are presented (Box 7.5). For example, consider forecasts of whether there will be rain: the outcome is discrete – there will either be rain or no rain. A discrete deterministic forecast (it will rain or it will not rain) can be scored as correct or incorrect, but such a scoring cannot be applied to a probabilistic forecast (e.g., there is a 20% chance of rain) without making important assumptions. Here we discuss the verification of only the most common situations.

Verifying deterministic forecasts

Deterministic forecasts can be measured for *accuracy* if the outcome is continuous, or for *correctness* if the outcome is discrete. A forecast is *accurate* if the forecast value is close to the observed value. *Precision* and *accuracy* are often confused or used synonymously. Precision is presumed accuracy: a precise forecast is one that gives an impression of being highly accurate, but may or may not be so. For example, a forecast of 20.1 °C implies that the temperature will be slightly more than 20.0 °C; whether that forecast is accurate is a separate question. Accuracy, correctness and precision are examples of *attributes* of forecast quality.

Inaccurate forecasts may be *biased*. Forecasts are biased if they more frequently overestimate or underestimate the actual weather. Bias is also indicated if the magnitude of errors of one sign are typically larger than those of the opposite. For example, forecasts of temperatures that are frequently too low have a cold bias. Similarly, deterministic forecasts of rain v no-rain are biased if rain is forecast too frequently or infrequently.

Accurate forecasts, or forecasts that are often correct, are not necessarily good forecasts.[21] For example, it is easy enough to make accurate forecasts of rainfall in deserts simply by forecasting no-rain all the time. The same problem applies with forecasts of any rare event: if you always forecast that the event will not happen you will nearly always be correct. A second forecaster who does occasionally predict some rain, may have fewer correct forecasts or a lower accuracy, but could still be producing more useful forecasts than the person who always predicts no-rain. For similar reasons, comparing accuracy for different locations or times of year can be misleading. A comparison of the accuracy of rainfall forecasts for Arica, Chile (the driest city in the driest country on Earth), with those for Buenaventura, Colombia (the wettest city in the wettest country on Earth) tells us more about the different climatologies than it does about the quality of the forecasts.

Verifying interval forecasts

To represent forecast uncertainty one option is to use an interval forecast, which consists of an upper and a lower limit between which a future value is expected to lie with a prescribed probability. *Reliability* is a critical attribute of good interval forecasts – does the observed value fall within the interval the correct number of times? If a 70% prediction interval is used, 70% of the forecasts should capture the observed value, and the observed value should fall outside the interval for 30% of the forecasts. If the observed value does not fall above and below the interval an approximately equal number of times, the forecasts are biased (cf. the definition of bias for deterministic forecasts).

The observed value should fall outside of the interval sometimes. If the observed value falls outside of the interval too frequently then the forecasts are overconfident – the interval is too narrow. Conversely, if the observed value falls outside of the interval too infrequently then the forecasts are under-confident – the interval is too wide. Because of the way reliability is measured, it is not possible to verify a single interval forecast: the one observation will either fall within the interval or outside it, and we cannot assess whether the proportion of intervals that contain the observation is correct.

It is possible to guarantee reliability by always issuing the same forecast and using knowledge of the climatology (§ 4.3.1), or even by cheating (issuing 70% of the forecasts with ridiculously wide intervals, and 30% with impossibly narrow or with unrealistic extremes). Therefore reliability is only one important attribute. However, there has been surprisingly little research on measuring other important attributes of interval forecasts.

Verifying probabilistic forecasts

As for interval forecasts, *reliability* is a critical attribute of good probabilistic forecasts. Similar problems apply – reliability can also be achieved from unhelpful forecasts, and individual probabilistic forecasts cannot be meaningfully verified – but there is at least a wealth of research on what additional attributes are important for probabilistic forecasts. Imagine that we are predicting rainfall occurrence. If rainfall occurs more frequently when the probability is high compared to when it is low, then the forecasts have *resolution* (not to be confused with spatial or temporal resolution). Alternatively, if the forecasts indicate higher probabilities when rain occurs compared to when it is dry, then the forecasts have *discrimination*.

How can I distinguish good from bad forecasts, or identify the best forecasts?

Skill is an attribute that can be applied to all types of forecasts. Forecasts are *skilful* when they outscore an alternative (usually, but not necessarily simple) set of forecasts. For deterministic forecasts, the alternative forecasts may be persistence (the assumption that the latest observation, or possibly the latest observed anomaly, will remain unchanged), the climatological average (§ 4.3), some other unchanging value (such as always forecasting no-rain in a desert) or random values (but with a realistic climatology). Forecasts may be skilful against one of these alternative forecasts, but not against another of them. It is important to examine the validity of the alternative forecasts carefully, since there are many ways of making forecasts look more skilful than they are useful. For probabilistic forecasts, the alternative forecasts are usually climatological probabilities, and the assumption is that if forecasts cannot outscore the climatology then it is best not to use the forecasts. Depending on the score that is used, that assumption is not always valid.[22]

BOX 7.5 FORECAST FORMATS

Weather and climate forecasts generally take one of various formats:

- *Deterministic*: a forecast with no accompanying information about the associated uncertainty. For example: the maximum temperature tomorrow will be 27 °C. This forecast indicates the expected temperature, but provides no information regarding the expected accuracy; it is not possible to calculate the chance of exceeding 30 °C, for example. Deterministic formats are commonly used for temperature forecasts at short- and medium-range weather timescales. The level of accuracy usually has to be learned by experience.
- *Ranges and prediction intervals*: an upper and a lower value between which the observed value is expected to occur. For example: rainfall accumulations of between 25 and 40 mm overnight. The range could be the maximum and minimum possible values, or, more likely, may be a prediction interval. For a prediction interval there is an implied probability that the observed value will lie within the interval. That probability is often 70%, but alternative probabilities are used. Prediction intervals are commonly used for rainfall forecasts at short- and medium-range weather timescales.
- *Probabilistic*: a forecast of the probability of occurrence of one or more discrete events. For example: there is a 60% chance of rain tomorrow. Probabilities are used frequently in short-range weather forecasting to indicate the chance that some rainfall will occur (so-called *probability of precipitation*, PoP, forecasts). Probabilities are also used extensively in seasonal forecasting (see Box 8.3). Occasionally, the probabilities may be expressed as *odds*. Odds are used in preference to probabilities in contexts such as risk assessments and sporting events. Odds are defined as the probability of a specific outcome occurring divided by the probability of it not occurring.
- *Multiple alternatives*: a forecast that shows all, or a selection of, the individual predictions. This option is applied when showing a collection of forecasts of El Niño / La Niña (see Box 5.1) from various models (Chapter 8). It is less often applied for weather forecasts, but the individual trajectories of storm tracks may be shown as an alternative to an idealized plume.
- *Full probability distributions* (typically shown as a probability of exceedance curve): a forecast that shows the probability of exceeding (or not exceeding) all possible thresholds across a range of values. Exceedance probability curves are used in flood forecasting,[23] and have had some limited application in seasonal forecasts.[24]

7.3 Why is it so hard to predict the weather beyond a few days?

Weather forecasts are made by trying to predict how the weather that is occurring now (the *initial conditions*) will change over the next few hours to days. Anticipating how the weather will change is actually the easier part of the problem; the more difficult problem is to know exactly what the weather is like now. The primary reason why weather forecasts sometimes fail is that the forecasters do not always have an adequate estimate of what the current weather is like.

The difficulty with forecasting the weather is that it is chaotic.[25] A chaotic system is one in which future states are highly sensitive to small differences in current states[26] – i.e., the smallest of difference now can evolve into substantial differences in the future. If two virtually indistinguishable starting values can give different outcomes, in effect there are different solutions to the governing equations; and, if there is more than one solution, we cannot be certain what the actual outcome will be.[27] The implication is that it is impossible to make accurate weather predictions in practice because even a slightly inaccurate estimate of the current weather will translate into a large error in the forecast at some stage in the future. Lorenz first articulated chaos theory when he observed that his meteorological model gave completely different results if he ran it using three decimal places instead of six. He called this extreme sensitivity the 'butterfly effect' because, taking the argument to its limit, it may be possible for the flapping of butterfly wings to cause a tornado half way around the world.[28]

The Earth's weather is chaotic, and therefore impossible to predict perfectly, but there are additional reasons why weather forecasts are inexact. We do not know perfectly all the equations that govern our weather, and the ones that we do know we cannot model exactly because of computer limitations. Weather forecasts are time-sensitive and so have to be made as quickly as possible, which means that the forecasters have to simplify their models. Even if the current weather were measured exactly, the forecasts would fail because of these simplifications. Having access to immensely powerful computers would still not be enough: there are random effects, such as volcanoes and various human activities, which again make perfect forecasting impossible. Despite all these theoretical and practical difficulties in forecasting the weather, a chaotic system may be predicted with reasonable accuracy, as long as we do not try to predict too far into the future. Useful forecasts of a chaotic system can also be made by indicating how sensitive the forecasts are, perhaps by communicating a set of possible outcomes rather than predicting one specific outcome.

7.4 Given that it is hard, how do forecasters make predictions?

To predict the weather, forecasters need to know: a) as much as possible about the current weather – all over the globe, and not just at the surface; b) how the weather patterns will change; and c) what those changes mean for the weather where people need forecasts. To predict the movement of storms over the next few minutes or hours, for example, forecasters can use simple extrapolation procedures, known as nowcasting techniques (Box 7.1), if good observations (perhaps from weather radars) are available. However, to predict accurately beyond about two hours, it becomes necessary to model how those observed weather systems will change. To make such predictions the following steps are taken:

- *Observation*: take measurements of the current state of the weather.
- *Analysis*: estimate the current state of the atmosphere over the whole globe and from the surface up to at least about 10 km, including where we do not have any measurements.
- *Initialization*: input these measurements and estimates into a computer model.
- *Integration:* use the model to predict how the current state of the atmosphere will evolve.
- *Post-processing*: determine how that future state will affect the weather at the locations and times of interest.

These steps are described in more detail in the following sub-sections. Case Study 7.3 illustrates how such techniques can be used in disease forecasting.

CASE STUDY 7.3 WEATHER FORECASTING TECHNIQUES FOR FLU FORECASTING

Jeffrey Shaman, Mailman School of Public Health, Columbia University, New York, USA

As for the atmosphere, the processes describing the propagation of infectious agents within a human or animal population, or between vectors of disease and hosts, are nonlinear and can be modelled mathematically. Indeed, to simulate an outbreak of a novel or recurrent infection such dynamical representation of the infectious disease system is imperative. Although the system is nonlinear, the methods used for numerical weather prediction can similarly be brought to bear to construct accurate, calibrated infectious disease forecasting systems.

Three principal 'ingredients' are needed to develop a dynamic infectious disease forecasting system. The first is a mathematical model describing the transmission of a particular pathogen through a population. Numerous model constructs exist for simulating disease transmission and range from simple, low-dimension compartmental models, which bin individuals by infection status, to higher dimension network model structures and agent-based forms, which represent individuals in the system. The second required ingredient is observation of the system itself. Observations are vital for optimizing the model prior to forecast – in essence, for training the model to represent conditions and transmission activity as thus far observed. For infectious disease systems, observations are typically near real-time estimates of infection incidence, which are derived from sentinel clinical networks. The third required ingredient is a Bayesian inference algorithm, or data assimilation method, which is used to carry out the model optimization in the presence of observations.

For the purposes of forecast, a low dimension model construct is often preferred as this enables better model optimization given limited observations. As an example, early forecasts of influenza and influenza-like illness were developed using a simple compartmental model, observations of incidence and a data assimilation method.[29–31] These efforts have generated accurate prediction of outbreak onset (the week at which influenza incidence rises above some baseline level), outbreak peak timing and magnitude, and total cases. Due to the weaker nonlinear dynamics governing disease transmission, accurate forecasting of influenza is possible two to three months in advance of an outcome. These long lead-times suggest that influenza forecasts will have considerable utility as they become integrated into real-time public health and medical decision-making. It should be noted that, as for weather, prediction of an infectious disease event prior to any observed activity is presently not feasible. That is, much as the specific landfall location and timing of a hurricane cannot be accurately forecast before a nascent tropical disturbance has been observed, specific local influenza outbreak outcomes cannot be predicted before incipient influenza activity has been observed. As a consequence, prediction of the emergence of a novel pathogen is not suited for the methods described here.

Infectious disease forecasts, like weather forecasts, are probabilistic and provide a distribution of potential future outcomes, which can be calibrated to provide the end-user still greater information. For example, a weather forecast does not merely indicate that rainfall will or will not occur; rather, predictions are provided as calibrated probabilities. With this calibration, precipitation occurs on roughly 80% of days for which an 80% chance of precipitation tomorrow has been predicted and on 20% of days for which a 20% chance of precipitation tomorrow has been predicted (see Box 7.4). Similar calibrated probabilities can be developed for infectious disease forecasts. Such discrimination of *expected* forecast accuracy provides end users with richer information. For example, a calibrated forecast of a 70% chance that influenza incidence will peak in five weeks has much more urgency than a forecast of a 10% chance that influenza incidence will peak in five weeks. Both forecasts predict the same outcome, but the former ascribes a much higher probability to the event occurring and may be actionable. In contrast, the latter indicates the best estimate is a peak in five weeks, but the likelihood of this event is low and the uncertainty high.

In the last five years, the field of influenza and infectious disease forecasting has advanced considerably. Influenza forecasts have been operationalized and delivered in real-time,[32,33] purely statistical forecasting approaches have been developed,[34] and ensemble forecasts have been generated.[35] The field is still in its infancy; however, with continued investment, one can expect to witness improved accuracy and application of infectious disease forecasts in ways not yet anticipated.

7.4.1 Observation

Observations of the weather are drawn from many sources:

- Surface observations of temperature, rainfall, winds, humidity, air pressure, etc. (§ 4.2). These observations are taken at weather stations over the land, and from ships and buoys in the sea. There are many formally designated weather stations that take measurements at set times of the day that are coordinated across the whole globe. The measurements are distributed electronically by international agreement between National Meteorological Services via the Global Telecommunications System (GTS) of the WMO (§ 6.2).
- Observations of temperature, air pressure, winds, and humidity, at different altitudes above the weather stations. These *upper-air soundings* are obtained by fastening an automatic weather instrument to a balloon, and transmitting the measurements by radio waves back to the weather station. The instruments are called *radiosondes* or *rawinsondes.*
- Observations from aircraft to supplement the upper-air soundings. These observations may be taken directly from the aircraft, or an instrument similar to the radiosonde, called a *dropsonde*, may be dropped and the observations transmitted back to the aircraft. Observations are taken routinely by air traffic, but special reconnaissance flights may be sent into major storms to take more detailed observations that assist with predicting the storms evolution. These reconnaissance observations have made important contributions to improvements in the accuracy of forecasts of tropical cyclones, for example.
- Remotely-sensed observations, such as satellite and radar measurements, can provide more complete spatial coverage than is possible from direct measurement (see § 6.3.2).

7.4.2 Analysis

The various observations are checked for likely errors based, in part, on consistency with nearby and preceding observations and with other parameters (for example, is the relative humidity physically consistent with the air temperature?). This quality-checking involves comparisons of the observations with the prediction made a few (typically six) hours earlier using a numerical weather prediction (NWP) model (see § 7.4.4 for further details on NWP models). The new observations that are accepted as reasonable are used to correct the previous forecast, and thus provide an estimate of the current state of the atmosphere. This estimate is called the *analysis*.

The upper-air data from the balloons and aircraft, and especially from the satellites, are by far the most important data used in the analysis. Only limited use is made of the surface observations, in part because observations near the ground can change substantially over just a few metres or within a few seconds, and so are often insufficiently representative of the surrounding area. However, the movement and evolution of storms and other weather systems are most strongly affected by winds a few kilometres above the surface, including the jet streams (§ 5.3.2); thus, upper-air soundings may be more important than surface observations.

The analysis is a very complex and computationally intensive step, and is performed only by those forecasting centres with the most powerful computers (§ 6.3.3). These analyses are made available to other centres, which, in turn, use them to initialize their own weather prediction models.

7.4.3 Initialization

Initializing a model involves specifying the starting values in the various equations in the model; thus, the equations in the NWP model need to be initialized with estimates of the current weather – the *initial conditions*. The NWP models' initial values cannot always be taken directly from the analysis because the NWP model may have a different grid or use different equations than the analysis model (see § 7.4.2). In these cases the analysis has to be interpolated to the NWP model grid, and possibly adjusted. Even if the same model is used, initialization is a distinct step when generating alternative initial conditions as a means of producing a set (an *ensemble*) of forecasts (as discussed in the following sub-section on Integration).

7.4.4 Integration

NWP models involve a set of equations, known as *primitive equations*, which are based on the physics of how air behaves. These equations describe what is happening at a specific location, and are calculated at many points, vertically and horizontally, over a region of interest or over the whole globe. Since 2016, the most complex NWP model has almost one billion gridpoints (about 6.5 million spread across the globe at each of 137 vertical layers). The equations are used to predict the values of various meteorological parameters from current values.

The primitive equations cannot be used to make an immediate prediction for 24 hours hence; instead they predict only a few minutes into the future. The equations are then updated using these new values to predict the subsequent few minutes ahead, and the process is repeated until the predictions extend as far into the future as desired. For example, if the equations are set to predict ten minutes ahead, then to predict one day ahead requires that the calculations are repeated 144 (6 per hour × 24 hours) times. The smaller this time step is, the more precise the equations are, but the longer the forecast will take to compute. Similarly, the equations work best if the distances between the gridpoints are small, but the forecast will again take a long time to compute if there are many gridpoints. Compromises have to be made on the model time step and the grid resolution (i.e., the number of gridpoints) to minimize computation time. The compromises are one set of reasons why forecasts are not always accurate.

A further complication is that some of the important processes cannot be represented by primitive equations, perhaps because they occur at spatial scales that are too small for the model to capture. Many clouds, for example, are much smaller than the grid-spacing of the model, and so many of the important mechanisms in the formation of rain cannot be modelled properly. These processes have to be simplified or approximated using alternative equations called *parameterizations*. Parameterizations are another reason why forecasts are not always accurate.

We have discussed three main sources of error in weather forecasts: 1) the initial conditions are imperfect; 2) the model is imperfect; and 3) the weather is inherently chaotic (making the first two imperfections real problems!). Given that forecasters know these sources, different strategies are available to try to address the problems:

- *Minimise the imperfections* by generating the best analysis possible, and using the best-possible NWP model to produce a single best-possible forecast. This forecast is known as a *deterministic* forecast. A good example of this approach is the European Centre for Medium Range Weather Forecasts (ECMWF), which is widely acknowledged to have one of the best weather forecasting systems in the world. However, since ECMWF produces only medium- and longer-range forecasts (Box 7.1), it is able to take more time than other centres to produce its analysis (waiting for additional observations to become available, and taking longer to perform a more careful analysis), and to run a more complex model.
- *Assess how much the imperfections matter* by generating multiple forecasts using slightly different estimates of the current weather, and/or by using different models (Box 7.6). If the individual forecasts in this *ensemble* do not differ much then we can presumably be confident that the imperfections do not matter very much, but if they do differ then the likely outcome is unclear. Ensemble forecasting has become standard for forecasts beyond about three days, and has enabled us to predict weather about 50% further into the future. Further details of how the ensemble is generated, and how the multiple predictions can be combined to produce a meaningful forecast are provided in Box 7.6.

BOX 7.6 ENSEMBLES

For all weather and climate forecasts beyond about three days, it is standard practice to generate multiple predictions. The collection of predictions for the same target period is called an 'ensemble'. The objective when generating an ensemble is to produce alternative predictions that account for different sources of uncertainty: the initial conditions and the model imperfections in the case of weather forecasts. Alternative ways of generating an ensemble of predictions are described here; some of these options are more suitable for weather forecasting, others for climate forecasting, as indicated.

Generating an ensemble from different initial conditions

Recognizing that the initial conditions in our NWP model are inexact, we could produce additional predictions using alternative (and ideally equally reasonable) initial conditions.

Perhaps the simplest way to generate an ensemble of predictions from different initial conditions is to use recent predictions as well as the latest one(s). For example, when predicting the weather for Tuesday, say, we could consider both the forecast for Tuesday that was made today as well as the forecast for Tuesday that was made yesterday. This combination of current predictions with slightly out-of-date predictions is known as *lagged-averaging*. The procedure is not generally used in weather forecasting because forecast accuracy decreases rapidly with increasing lead-time: yesterday's forecast for Tuesday is likely to be less accurate than today's forecast for Tuesday. However, lagged-averaging is used in long-range (e.g., seasonal) forecasting (Table 7.1) where the initial conditions are no longer the primary basis for prediction (see § 8.2.2).

For extended- and shorter-range forecasts, the older predictions are too inaccurate. An alternative option is to take the most recent prediction, called the control run, and to make additional predictions by perturbing the initial conditions from the analysis. The perturbations need to be introduced in ways that organize into alternative weather patterns. Defining these *dynamically constrained perturbations* involves complex statistical procedures that identify where past errors in the forecasts have grown most rapidly. There are various ways of identifying such perturbations, and one or more of these options is the preferred approach for generating ensembles by the leading weather forecasting centres around the world.

An additional possible option is to use alternative analyses in the model initialization. This procedure is widely used with regional NWP models, but it is less often used in global models. The analysis produced by one global NWP model may result in an initialization that is numerically unstable in a second global model. The effect is that the second model quickly predicts highly unrealistic weather conditions, or the model may even crash.

Generating an ensemble from different or modified models

A conceptually simple way of generating an ensemble is to use forecasts from different NWP models. This approach is used widely for extended and long-range forecasts; forecasts produced by different centres are combined into a *multi-model* ensemble.[36,37] For shorter-range forecasts, there may be practical and other reasons why an ensemble might need to be developed in-house, but maintaining the computer code for different models in an operational setting is expensive. A cheaper option is to modify a single NWP model's simplified equations – the parameterizations (§ 7.4.4). These parameterizations can be modified in two ways: by adding random errors at each calculation to represent the uncertainty in the parameterizations (so-called *stochastic physics*); or by using an alternative parameterization across every calculation (so-called *perturbed parameterization*). These methods have become standard for medium-range forecasting.

7.4.5 Post-processing

The outputs from the NWP model(s) may be inadequate as a forecast for various reasons:

- The model produces predictions at its gridpoints, which may not be in the locations that forecasts are needed.
- The model outputs represent area-averages, which can be misleading, especially in the case of rainfall. Rainfall is often highly localized (§ 5.2.5.2), but the model cannot produce rainfall events smaller than its grids, and so localized downpours will instead be represented as widespread drizzle over the whole grid. As a result, the frequency of rainfall is likely to be overestimated, but the intensity underestimated. Similarly, the area-average temperature may not be representative of the location of interest, especially in mountainous areas.
- The models can have large systematic errors, such that temperature forecasts, for example,[38] may be consistently too high or too low.

Each of these problems can be addressed by applying a statistical correction to the NWP model output. The statistical correction is applied by comparing past forecasts with observations. A forecast that is derived from NWP output and then corrected using a statistical model is called *model output statistics* (MOS) guidance. The MOS techniques can range from simple linear regression to highly complex procedures using ensembles of forecasts from more than one NWP model.

7.5 How accurate are weather forecasts?

There is a common tendency to see weather forecasts as less accurate than they really are because of implicit expectations that forecasters should be able to predict exactly *when* a *specific* weather event will occur. Imagine that on a Friday a meteorologist predicts that there will be a storm bringing heavy rain early the following week. The forecaster is not sure precisely when the storm will arrive but the best estimate is Monday afternoon. The storm approaches more slowly than anticipated and only arrives Tuesday morning. As a result, Monday is sunny when it was forecast to be wet, while Tuesday is wet when it was forecast to be sunny, and so the forecast looks to be wrong twice. If a health worker postpones a visit to an isolated village on Monday afternoon to Tuesday morning to avoid the storm she may be justifiably annoyed. However, it does seem unfair to blame the forecaster who correctly predicted that a storm was approaching. If the health worker had known that the exact timing of the storm was uncertain, she may have been able to postpone her visit until the Wednesday or Thursday, well after the storm should have passed by.

The forecaster, while correct in predicting a storm, could have provided more useful information by indicating the uncertainty in the timing of the storm (Box 7.1). Of course, it would be wonderful to know exactly when the next storm will occur; just as it would be helpful for a midwife to know exactly when an expectant mother will go into labour. But neither are perfectly predictable. However, it is still possible to plan for a storm, or a birth, with only a rough sense of when it will occur – anytime in the next three to five days, perhaps. As the storm becomes more imminent, it may be possible to get a more precise forecast of where it will hit when, and how intensely.

One further complication is that measurements of the quality of forecasts need to be targeted at forecasts of the specific weather conditions of interest. For example, knowing that temperature forecasts for your city are, on average, accurate to within 0.5 °C, may be misleading if the most accurate forecasts are for all the days when temperature is close to average but one is only interested in the forecasts of extreme temperatures.

Accounting for all such intricacies, and many others besides, makes evaluating forecasts in a scientifically rigorous, but intuitive, way an almost impossible task (Box 7.4). Given that describing how good (or bad) weather forecasts are is a far more complex question than it may at first appear, it is possible here to provide only some crude generalizations. Perhaps the simplest place to start is with how weather forecasts have been improving with time. A simple rule of thumb is that since at least the 1970s, the lead-time of a forecast for a given level of accuracy has improved by about one day each decade. In other words, three-day forecasts that are available now are as accurate as the two-day forecasts that were available ten years ago. As a second simple rule of thumb, one week is about the limit for usable deterministic forecasts of individual weather systems; that limit extends to about ten days for probabilistic forecasts. There are attempts to predict weather conditions beyond this ten-day limit, in so-called sub-seasonal forecasts (Box 7.4), but forecast products remain primarily research initiatives. Forecasts at such long lead-times may be available from some commercial centres, but many commercial weather forecasts do not provide adequate estimates of skill and can be misleading.

7.5.1 Temperature

Temperature is generally easier to predict than rainfall, largely because of the highly localized nature of rainfall (§ 5.2.5). Figure 7.1 shows a measure of our ability to predict hot and cold spells three days in advance. The two main features are:

- Cold spells are predicted more successfully than hot spells. The cold spells are easier to predict because they are generally associated with large-scale wind patterns, possibly in conjunction with cold fronts (§ 4.2.8.2). Hot spells, in contrast, can be sensitive to small differences in cloudiness and to surface conditions for which observational data may be limited. However, although cold spells can generally be predicted more successfully than hot spells, summer temperatures (in the mid-latitudes) are typically more accurate than winter temperatures simply because summer temperatures are less variable (§ 5.3). This apparent paradox is partly an effect of the distinction between accurate and skilful forecasts (Box 7.4).
- Forecasts are better in the mid-latitudes than in the tropics. Mid-latitude temperature forecasts are relatively easy for similar reasons to the cold / hot spell contrast: they are determined by large-scale wind patterns (§ 5.2.5). In the tropics temperatures are more uniform and unvarying, and days that qualify as part of a hot spell can be highly localized depending on small differences in cloudiness or surface heating.

7.5.2 Rainfall

Forecast probabilities of rainfall occurrence are generally reliable (Box 7.4) for all weather timescales if the NWP model outputs are post-processed (§ 7.4.5). What changes with the increasing lead-time is that fewer probabilities of close to 0% or 100% are issued. If forecast probabilities are reliable, the quality of the forecasts is essentially built in to the forecast itself, and there may be no need for further verification information. When the forecast says there is a 60% chance of rain, the probability indicates how often the warning is a correct alarm if the forecast is interpreted as a warning of rain.

Predicting rainfall amounts is difficult, especially for convective rainfall (§ 4.2.2) because it is so localized (which makes predicting its occurrence difficult too). The smallest useful scale for 24-hour forecasts of rainfall is about 85 km.[40] Largely because convective rainfall is harder to predict than large-scale rainfall, forecasts in the tropics are not as good as forecasts in the extra-tropics. For the same reason, summer rainfall is harder to predict than winter rainfall.

7.5.3 Tropical storms (cyclones, hurricanes and typhoons)

Storm-track forecasts have steadily improved over the last few decades as a result of the ever-increasing quantity and accuracy of atmospheric observations,

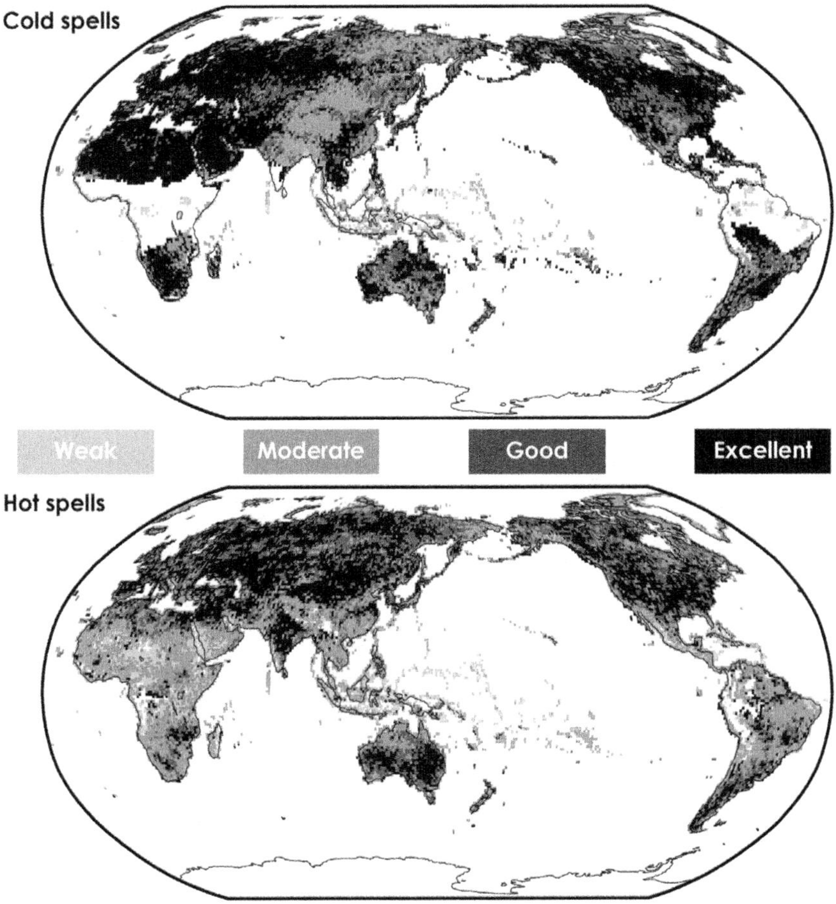

FIGURE 7.1 How well can we forecast severe heat or cold? Ability to discriminate (see Box 7.2) cold (top) and hot (bottom) spells from all other days. The spells are defined as temperatures beyond the 5th and 95th percentiles, respectively, on at least two consecutive days. The forecasts are derived from ECMWF's ensemble prediction system (see § 7.4.4), and are verified against the ECMWF analysis (which means that the quality of the forecasts are over-estimated). The score is classified as 'excellent' (> 95%), 'good' (90–95%), 'moderate' (67–90%) or 'weak (< 67%), and guessing would score 50%. Some areas show no score because of insufficient numbers of spells. (Data source: adapted from Coughlan de Perez et al.[39])

improvements in the way those observations are used to initialize NWP models, improvements in the models themselves and the benefits of access to more powerful computers. The tracks of tropical cyclones are predicted most accurately in the North Atlantic because these storms are monitored carefully and so the prediction models can be initialized most accurately. A sample of average errors for the 2016 storms is shown in Table 7.3, together with previous errors to show

TABLE 7.3 Approximate average errors (in km) in predicting North Atlantic tropical storm tracks

Lead-time	*2016*	*1996*	*1976*
24 hours (1 day)	50	135	240
48 hours (2 days)	100	240	530
72 hours (3 days)	155	350	750
96 hours (4 days)	240	N/A	N/A
120 hours (5 days)	330	N/A	N/A

Source: National Hurricane Center Forecast Verification Official error trends[ii]

improvements over the last 40 years. Average errors do fluctuate from year-to-year by about 20–50%, and errors in other ocean basins are generally no more than about 20% larger than those shown in the table. Most tropical cyclones are less than about 1000 km in diameter (§ 4.2.8), so the errors for five-day forecasts are about one-third the size of a large hurricane. Whether such errors are considered large or small depends on one's perspective. Failing to act and acting in vain have to be weighed carefully by decision-makers (§ 7.3).

7.6 What weather forecasts are available?

7.6.1 Watches and warnings of hazardous and inhospitable conditions

When there is a risk of severe weather conditions, most countries provide official alerts. For many reasons, it is important to work with alerts from mandated government authorities (typically the NMHS), even if there is a perception that higher quality information may be available elsewhere. There are two levels of alert:

- *Watches*: indicate that hazardous conditions are *possible*. During a watch the appropriate response is to prepare for action should the hazardous weather strike, and to keep a look out for updated information.
- *Warnings*: indicate that hazardous conditions are *expected*. During a warning the appropriate response is to take action immediately.

In the event of a watch or a warning, some countries may have standard procedures that relevant government agencies are expected to follow. However, sometimes it is useful to receive notification of hazardous weather that does not meet the criteria for an alert. The NMHS may then issue an *advisory*, which does not have the official status of an alert, but which may still warrant action. Each country has its own criteria for issuing advisories and alerts, and its own protocols to follow for disseminating the information, so there may be some apparent inconsistencies across borders. The WMO collates the various active official alerts in its Severe Weather Information Centre.[iii] MeteoAlarm[iv] is a more detailed site specifically for Europe.

As part of the World Weather Watch (§ 6.2), the NMHSs are supported by a set of Regional Specialized Meteorological Centres, which are responsible for monitoring specific hazards. There are centres to monitor tropical cyclones, nuclear accidents, volcanic ash, wildfires, floods and droughts.

Many NMHSs are now developing as multi-hazard warning centres because they already provide a 24/7 service and are mandated authorities for weather warnings.[41] As an example, heat health early warning systems are being established in many countries through collaboration between NMHSs and public health departments. As a measure of heat-related health risk, temperature alone is inadequate: it is the combination of unusually high temperatures with high humidity that constitutes a health threat (see Box 4.2). Defining a heat index that is applicable everywhere is problematic, and even for a given location, the health impacts of heat stress vary through the year. In the mid-latitudes, for example, humans are more sensitive to heat stress in early summer even though the heat index typically reaches peak values a few weeks or even months later. Similarly, for cold spells, it is the combination of unusually low temperatures with windy conditions, rather than temperatures alone, that pose the greatest threat. Because of these compounding effects of humidity and wind and of human adaptability to location and time of year, heat wave and cold spell monitoring and early warning systems have to be adapted to the local environment.

7.6.2 Forecasts of unhealthy weather

In some cases it may be important to include air quality measurements in the heat health early warnings. For example, increased near-surface ozone concentrations are often a problem in heat waves, while in winter certain weather conditions can be conducive to both extremely cold conditions and increased air pollution (§ 5.2.4; see also § 6.4.2.2 on air quality monitoring). The Regional Specialized Meteorological Centres (§ 7.6.1) are responsible for providing forecasts and information on additional transboundary unhealthy weather conditions, such as dust and radioactive material.

7.6.3 Forecasts of suitable weather

Weather forecasts that indicate the timing of a seasonal window for disease occurrence may also be useful in health decision-making. While weather forecasts may only give a few days' indication of the start of the normal transmission window for a seasonal respiratory disease, the prediction that the season is nearing its end allows health decision-makers to scale back their response and re-focus their attention elsewhere. For example, meningococcal meningitis epidemics occur during the hot and dusty dry season in the African Sahel when relative humidity is low (see Case Study 7.1). A weekly average humidity below 40% is predictive of epidemic occurrence. Weather forecasts that predict a rise in relative humidity above 40% and the end of the seasonal epidemic window have been developed to help health practitioners plan the end of their seasonal vaccination activities.[7]

7.7 Conclusions

Weather forecasts are more accurate than many people suppose them to be, but predicting rainfall accurately is notoriously difficult, especially in the tropics. Despite their accuracy, weather forecasts may not add much value to a health early warning system if there is a long lead-time between the observed weather and the health outcome; e.g., for vector-borne diseases. However, weather forecasts can be useful for other purposes, such as for prediction of hazardous conditions. If weather alerts of hazardous conditions are to be used, it is important to work with alerts from officially mandated sources. Weather forecasts can provide at most a few days' advanced notice; for informing preparedness actions weeks or months in advance, it may be possible to use seasonal forecasts – the topic of the following chapter.

Notes

i www.wmo.int/pages/prog/www/DPS/GDPS-Supplement5-AppI-4.html.
ii https://www.nhc.noaa.gov/verification/verify5.shtml.
iii http://severe.worldweather.org/.
iv www.meteoalarm.eu/.

References

1 Rogers, D. Partnering for health early warning systems. *World Meteorological Bulletin* **60**, 14–18 (2011).
2 Sarran, C., Halpin, D., Levy, M. L., Prigmore, S. & Sachon, P. A retrospective study of the impact of a telephone alert service (Healthy Outlook) on hospital admissions for patients with chronic obstructive pulmonary disease. *Primary Care Respiratory Medicine* **24**, 14080, doi:10.1038/npjpcrm.2014.80 (2014).
3 Tang, X., Feng, L., Zou, Y. & Mu, H. in *Institutional Partnerships in Multi-Hazard Early Warning Systems: A Compilation of Seven National Good Practices and Guiding Principles* (ed. M. Golnaraghi) (Springer, Heidelberg, 2012).
4 Thiery, W. *et al.* Early warnings of hazardous thunderstorms over Lake Victoria. *Environmental Research Letters* **12**, 074012, doi:10.1088/1748-9326/aa7521 (2017).
5 WMO. *Sand and Dust Warning and Advisory and Assessment System (SDS-WAS) Science and Implementation Plan 2015:2020.* 30pp (WMO, Geneva, 2014).
6 Thomson, M. C. *et al.* in *Priorities in Climate Research, Analysis and Prediction* (eds. J. Hurrell & G. Asrar), 459–484 (Springer, Dordrecht, 2013).
7 Pandya, R. *et al.* Using weather forecasts to help manage meningitis in the West African Sahel. *Bulletin American Meteorological Society* **96**, 103–115 (2014).
8 Perez Garcia-Pando, C. *et al.* Soil dust aerosols and wind as predictors of seasonal meningitis incidence in Niger. *Environmental Health Perspectives* **122**, 679–686 (2014).
9 Jusot, J.-F. *et al.* Airborne dust and high temperatures are risk factors for invasive bacterial disease. *The Journal of Allergy and Clinical Immunology in Practice* **139**, 977–986, doi:10.1016/j.jaci.2016.04.062 (2017).
10 Robine, J.-M. *et al.* Death toll exceeded 70,000 in Europe during the summer of 2003. *Comptes Rendus Biologies* **331**, 171–178 (2008).
11 Lowe, D., Ebi, K. L. & Forsberg, B. Heatwave early warning systems and adaptation advice to reduce human health consequences of heatwaves. *International Journal of Environmental Research and Public Health* **8**, 4623–4648, doi:10.3390/ijerph8124623 (2011).
12 Nissan, H., Burkart, K., Mason, S. J., Coughlan de Perez, E. & van Aalst, M. Defining and

predicting heat waves in Bangladesh. *Journal of Applied Meteorology and Climatology* **56**, 2653–2670, doi:10.1175/JAMC-D-17-0035.1 (2017).
13 Pascal, M. *et al.* France's heat health watch warning system. *International Journal of Biometeorology* **50**, 144–153 (2006).
14 Knowlton, K. *et al.* Development and implementation of South Asia's first heat-health action plan in Ahmedabad (Gujarat, India). *International Journal Environmental Research and Public Health* **11**, 3473–3492, doi:10.3390/ijerph110403473 (2014).
15 National Academies of Sciences, E., and Medicine. *Next Generation Earth System Prediction: Strategies for Subseasonal to Seasonal Forecasts.* 366pp (The National Academies Press, 2016).
16 Lowe, R. *et al.* Evaluating the performance of a climate-driven mortality model during heat waves and cold spells in Europe. *International Journal of Environmental Researh and Public Health* **12**, 1279–1294, doi:10.3390/ijerph120201279 (2015).
17 McKinnon, K. A., Rhines, A., Tingley, M. P. & Huybers, P. Long-lead predictions of eastern United States hot days from Pacific sea surface temperatures. *Nature Geosciences* **9**, 389–394, doi:10.1038/ngeo2687 (2016).
18 Conti, G. O., Heibati, B., Kloog, I., Fiore, M. & Ferrante, M. A review of AirQ Models and their applications for forecasting the air pollution health outcomes. *Environmental Sciece and Pollution Research* **24**, 6426–6445, doi:10.1007/s11356-016-8180-1 (2017).
19 Murphy, A. H. What is a good forecast? An essay on the nature of goodness in weather forecasting. *Weather and Forecasting* **8**, 281–293 (1993).
20 Jolliffe, I. T. & Stephenson, D. B. E. *Forecast Verification: A Practitioner's Guide in Atmospheric.* 2nd edn (Wiley, Chichester, 2012).
21 Murphy, A. H. The Finley affair: a signal event in the history of forecast verification. *Weather and Forecasting* **11**, 3–20 (1996).
22 Mason, S. J. Understanding forecast verification statistics. *Meteorological Applications* **15**, 31–40 (2008).
23 Piechota, T. C., Chiew, F. H., Dracup, J. A. & McMahon, T. A. Development of exceedance probability streamflow forecast. *Journal of Hydrologic Engineering* **6**, 20–28 (2001).
24 Barnston, A. G., He, Y. & Unger, D. A. A forecast product that maximizes utility for state-of-the-art seasonal climate prediction. *Bulletin of the American Meteorological Society* **81**, 1271–1279 (2000).
25 Lorenz, E. N. Deterministic nonperiodic flow. *Journal of Atmospheric Sciences* **20**, 130–141 (1963).
26 Smith, L. *Chaos: A Very Short Introduction.* Vol. 159 (Oxford University Press, Oxford, 2007).
27 Slingo, J. & Palmer, T. Uncertainty in weather and climate prediction. *Philosophical Transactions of the Royal Society A* **369**, 4751–4767 (2011).
28 Lorenz, E. N. Predictability: does the flap of a butterfly's wings in Brazil set off a tornado in Texas. *American Association for the Advancement of Science* **20**, 260–263, doi:http://gymportalen.dk/sites/lru.dk/files/lru/132_kap6_lorenz_artikel_the_butterfly_effect.pdf (1972).
29 Ong, J. B. S. *et al.* Real-time epidemic monitoring and forecasting of H1N1-2009 using influenza-like Illness from general practice and family doctor clinics in Singapore. *PLOS One* **5**, e10036, doi:10.1371/journal.pone.0010036 (2010).
30 Dukić, V., Lopes, H. F. & Polson, N. G. Tracking epidemics with google flu trends data and a state-space SEIR model. *Journal of the American Statistical Association* **107**, 1410–1426 (2012).
31 Shaman, J. & Karspeck, A. Forecasting seasonal outbreaks of influenza. *Proceedings of the National Academy of Sciences* **109**, 20425–20430, doi:10.1073/pnas.1208772109 (2012).
32 Shaman, J., Karspeck, A., Yang, W., Tamerius, J. & Lipsitch, M. Real-time influenza forecasts during the 2012–2013 season. *Nature Commentary* **4**, 2837 (2013).
33 Biggerstaff, M. Results from the centers for disease control and preventions predict the 2013–2014 Influenza Season Challenge. *BMC Infectious Diseases* **16**, 357 (2016).

34 Brooks, L. C., Farrow, D. C., Hyun, S., Tibshirani, R. J. & Rosenfeld, R. Flexible modeling of epidemics with an empirical Bayes framework. *PLOS Computational Biology* **11**, e1004382, doi:10.1371/journal.pcbi.1004382 (2015).
35 Yamana, T. K., Kandula, S. & Shaman, J. Superensemble forecasts of dengue outbreaks. *Journal of the Royal Society Interface* **13**, 123, doi:10.1098/rsif.2016.0410 (2016).
36 Palmer, T. N. *et al.* Development of a European ensemble system for seasonal to inter-annual prediction. *Bulletin of the American Meteorological Society* **85**, 853–872 (2004).
37 Barnston, A. G., Mason, S. J., Goddard, L., DeWitt, D. G. & Zebiak, S. E. Multimodel ensembling in seasonal climate forecasting at IRI. *Bulletin of the American Meteorological Society* **84**, 1783 (2003).
38 Clark, M. P. & Hay, L. E. Use of medium-range numerical weather prediction model output to produce forecasts of streamflow. *Journal of Hydrometeorology* **5**, 15–32 (2004).
39 Coughlan de Perez, E. *et al.* Global predictability of temperature extremes. *Environmental Research Letters* **13**, 054017 (2018).
40 Roberts, N. Assessing the spatial and temporal variation in the skill of precipitation forecasts from an NWP model. *Meteorological Applications* **15**, 163–169, doi:10.1002/met.57 (2008).
41 Golnaraghi, M. E. *Institutional Partnership in Multi-Hazard Early Warning Systems. A Compilation of Seven National Good Practices and Guiding Principles*. 350pp (Springer, London, 2012).

8

CLIMATE FORECASTS FOR EARLY WARNING

Up to six months in advance

Simon J. Mason
Contributors: Madeleine C. Thomson and Ángel G. Muñoz

> Summer shall come, and with her all delights,
> But dead-cold winter must inhabit here still.
>
> The Two Noble Kinsmen *by John Fletcher / William Shakespeare*

8.1 Introduction

Numerical weather prediction only became practically possible in the 1960s with the availability of sufficiently powerful computers. Prior to then, scientifically-based weather forecasting was based on statistical relationships informed by understanding of the physical processes of weather. Earlier still, weather lore,[1] some of which has some scientific basis,[2] has been used for centuries to make weather forecasts. However, the earliest record of a meteorological prediction is a seasonal climate forecast rather than a weather forecast. In the *Epic of Gilgamesh*, the god Ea warns Utnapishtim (more commonly known through the story of Noah, or Nûh ibn Lamech ibn Methuselah) of persistent torrential rain and a resulting flood. Utnapishtim successfully takes evasive action by constructing a boat (or ark) for himself and for the animals.

Warnings are only useful if they elicit an effective response, if they are clearly articulated and disseminated by mandated authorities, and are accepted as valid and actionable by the intended beneficiary community; budgets must be agreed, commodities moved to the area at risk, public communication campaigns developed, etc. (And, apparently, Utnapishtim was able to tick all those boxes.) All of these actions take time, and weather forecasts provide sufficient warning only to take limited action. If decision-makers could be provided with additional lead-time, they would be better able to organize an effective response.

In Chapter 7, we showed that in most cases it is not possible to make useful weather forecasts much beyond a week (unless we have the prophetic gift of an Utnapishtim). Despite this limitation, forecasts are routinely available at longer ranges. How can that be? Surely the further into the future one tries to predict, the less accurate the forecast will be. A simple response is that it is not true; sometimes it is easier to predict the further into the future one looks, just as a doctor can more accurately predict the status of a patient's cold two months hence (it will have passed) than its status in three days' time. What will differ across the various timescales of predictions are their accuracy and specificity (Chapter 3, Figure 3.1 and Box 8.1).

BOX 8.1 HOW DOES A SEASONAL CLIMATE FORECAST DIFFER FROM A WEATHER FORECAST?

A seasonal climate forecast is an indication of some aspect of the expected weather conditions aggregated over a period of between one and about six months, and typically starting a few weeks to a few months in the future. A typical seasonal climate forecast differs from a typical weather forecast in important ways (Table 8.1), although exceptions can be found. For example, while most seasonal forecasts are probabilistic, a few use intervals (e.g., predictions of tropical storm frequencies[3]).

TABLE 8.1 Differences between weather and seasonal climate forecasts

Characteristic	*Weather forecast*	*Seasonal forecast*
Specificity	Specific timing and intensity	General frequency and intensity
Parameters	Rainfall, max and min temperature, humidity, wind speed and direction, cloudiness	Rainfall, average temperature
Format	Deterministic; prediction intervals for rainfall	Probabilistic
Precision	Nearest °C; prediction intervals for rainfall and wind speed	3 categories (below, normal, above; Box 8.3)
Spatial resolution	Individual locations	Area-averages
Temporal resolution	Hourly to daily	3–4 months

TABLE 8.1 (Continued)

Characteristic	*Weather forecast*	*Seasonal forecast*
Skill (Box 7.4)	High	Moderate at best, often non-existent
Areas of greatest skill	Extratropics	Tropics
Source of skill	Initial conditions	Boundary conditions

The lack of skill in most seasonal forecasts is reflected in the forecast probabilities, which rarely differ from climatological probabilities by much. However, there are also some problems with the reliability of many seasonal forecasts[4] (§ 8.4), and so the probabilities may not provide a good indication of the uncertainty.

In this chapter, we explore the potential value of seasonal climate forecasts to the health community. We begin by considering why the general weather conditions over a season might be predictable, before examining how seasonal forecasts are made, and why they are presented in a different way to weather forecasts. We then examine where and when seasonal forecasts work best, and emphasize that it is possible to make useful seasonal forecasts for health outcomes only in some parts of the world and for certain times of the year (and possibly only for some years). We also review some of the main sources of seasonal forecasts.

8.2 How do forecasters predict the next few months?

When preparations for the 2014 World Cup soccer tournament were underway, concern was raised about the possibility of a dengue epidemic impacting the games, which were being held in cities across Brazil. A dengue early warning system was created that was driven by seasonal forecasts and predictions were made for each participating city.[5] How did this forecast differ from a prediction based on weather forecasts, and how was such a forecast even possible?

Consider a 'Spot the Ball' puzzle. Such puzzles are common in newspapers such as the *New York Times*.[i] They involve a photograph of a soccer (or other sports) match, but the ball has been removed. The problem is to estimate exactly where the ball should be. That question is analogous to estimating what the weather conditions are like at this moment, given the available observations. Now try estimating where the ball will be in ten seconds' time, or two minutes' time. We may be able to estimate where the ball will be in a few seconds' time if we know where it is now, but in two minutes' time the ball could be virtually anywhere, regardless of which is the better team. Although we cannot predict where the ball will be more than a few seconds into the future, it may well be possible to predict who will win the match, or who will win the tournament. For that, we need to know which is the better team. Analogously, estimating the current weather conditions (where the ball is now) is very difficult, and the weather is predictable only a short way into the future (where the ball will be in a few seconds' time). Nevertheless, it may be possible to predict the general weather conditions over the next few months without knowing what the weather will be like at any given time (similar to predicting the final outcome of the soccer match, but not when the goals will be scored).

Forecasts far beyond one week are sometimes possible if forecasters do not try to predict the weather at any specific time, but instead try to predict the general weather conditions over a prolonged period. For example, a forecast of generic weather conditions over the next few months might consider the question of whether there will be many storms – as opposed to, when will specific storms occur? In general, the further into the future the forecast is projected, the longer the period over which the predicted weather conditions are aggregated: typically about one-to-two weeks in the case of sub-seasonal forecasts (Box 7.2); one to four months in the case of seasonal forecasts (Box 8.1); or ten to 30 years in the case of longer-range projections and scenarios (Chapter 9). These aggregated weather conditions describe the climate (Box 4.1), and the reasons why scientists can predict the climate are not the same as the reasons why they can predict the weather. Similarly, the reasons why scientists can predict the climate over the next few months are not the same as why they can predict the climate decades into the future. Timescales of weather and climate variability, their causes and sources of uncertainty are described in Chapter 5, Table 5.1.

8.2.1 Why is the seasonal climate (sometimes) predictable?

As discussed in § 5.3.5, differences in climate from year-to-year can be substantial. An unusually wet season will occur if there is an excess number of rainfall-producing weather events, and/or they are more intense or persistent. Much of these year-to-year differences are completely random, and it is only possible to forecast the individual events at weather timescales. In some cases, however, there may be a reason why the weather behaves unusually. Perhaps the most clear-cut case is after a large volcanic eruption: large amounts of dust many kilometres up can block out the sun and so parts of the globe may cool down noticeably for possibly two years or even longer after the severest eruptions (§ 5.4.2.1). Such large volcanic eruptions are rare and unpredictable, and so their effects can only be predicted after the eruption has occurred. If seasonal forecasts are to be made more regularly than only after volcanic eruptions, other influences on the weather must be sought.

The key to predicting seasonal climate conditions was noted in Chapters 4 and 5: the air is heated by Earth's surface rather than directly by the sun, and so prolonged unusual conditions at the surface will have a (possibly predictable) effect on the climate. Earth's surface consists of sea and other water bodies, land and snow/ice, each of which provides some level of seasonal predictability.[6] Collectively, these surface conditions are called the *boundary forcings*. The mechanisms involved are discussed in the following subsections.

8.2.1.1 *The oceans*

Sea-surface temperatures are the most important source of predictability of seasonal climate[7,8] because:

- Most (70%) of Earth's surface is sea.
- Sea temperatures change much more slowly than air temperatures, and they can therefore have a prolonged effect on the weather.
- The main predictable effect on the air is through changes in humidity rather than changes in temperature, and the oceans are the primary source of moisture.

If the sea is unusually hot or cold, sea temperature anomalies may last weeks or months[9] because it takes so much energy to heat up and cool down water (§ 5.2.3). Air heats up and cools down much more easily than water, and so air temperatures adjust to sea temperatures much more quickly than vice versa. Therefore, sea-surface temperature anomalies can cause large and prolonged changes in evaporation and heating or cooling of the overlying air. Changes in evaporation are important not only because of how much water is available for making rain, but because of the latent heat in the water vapour (§ 4.2.8.1).[10]

The effect of sea-surface temperature anomalies is strongest in the tropics where the sea is hottest because the amount of moisture that the air can hold is more sensitive when the air is hot than when it is cold (§ 4.2.1). Therefore, a 1 ° increase in sea temperature in a hot sea can result in a much larger increase in evaporation than can a similar increase in sea temperature in a colder place. How then do sea-surface temperatures in different areas of the oceans affect climate?

8.2.1.1.1 Tropical Pacific Ocean

Variability in tropical Pacific sea-surface temperatures is dominated by the El Niño – Southern Oscillation (ENSO; Box 5.1). The ENSO is the main reason why seasonal forecasts are possible, because it has a stronger influence on temporal variability in climate at seasonal scales than anything other than the cycle of summer and winter (§ 5.3.3). The development of a numerical model in the 1980s that could predict ENSO events a few months in advance was based on the physics of how wind patterns and ocean currents in the equatorial Pacific Ocean affect each other.[11,12] The success of this model in forecasting the 1988 El Niño was a major stimulus for promoting widespread interest in operational seasonal forecasting, and the motivation for the creation of what is now the International Research Institute for Climate and Society (IRI).[13] Prior to then, forecasts were developed in only a handful of countries (§ 8.2.2).

Over the equatorial Pacific itself, areas of prolonged heavy rain preferentially occur over the warmest part of the ocean, and so these areas may shift thousands of kilometres between El Niño, neutral and La Niña episodes. Because Pacific Ocean sea temperatures near the equator are the highest in the world, the convective rainfall (§ 4.2.2) here is particularly heavy and widespread, such that the rainstorms are large and violent enough to affect weather patterns elsewhere. The ENSO can therefore affect climate in areas well beyond the equatorial Pacific, and so El Niño and La Niña are often used to predict seasonal climate anomalies in areas that do not even border the Pacific Ocean. These remote effects – or *teleconnections* – occur partly because weather patterns around the world respond to the shifts in weather patterns over the equatorial Pacific itself (for example, over southern parts of North America). Remote effects can also occur because some of these changes in weather patterns can disrupt wind patterns over the oceans, which changes the sea temperatures there (for example, over eastern and southern Africa because of changes in the tropical Indian Ocean.

8.2.1.1.2 Tropical Atlantic Ocean

The Atlantic Ocean does experience an El Niño-like phenomenon, called the Atlantic Equatorial Mode.[14] The warming events, Atlantic Niños, can cause drought in the Sahel and increased rainfall along the Gulf of Guinea. However, Atlantic Niños are able to develop only to about half the strength and persistence of those in the Pacific because the Atlantic is so much narrower.

Of greater importance for predicting seasonal climate than Atlantic Niños is the difference in temperature between the North and South Atlantic Ocean. Variations in the difference in sea-surface temperature across the tropical Atlantic have important implications for rainfall over much of West Africa and Northeast Brazil. On a larger-scale the north–south contrast in sea-surface temperatures throughout the Atlantic Ocean and beyond contribute to climate variability at decadal scales over areas such as West and North Africa and India.[15]

8.2.1.1.3 Tropical Indian Ocean

Sea-surface temperature variability in the Indian Ocean is weaker than in the Pacific and Atlantic Oceans, in part because of its size, but primarily because of the influence of the South Asian land-mass in the Northern Indian Ocean. The land-sea contrast and its alternation between summer and winter (§ 5.2.3) dominate the mechanisms of climate in the Indian Ocean, whereas in the Pacific and, to a lesser extent, the Atlantic, there is less interference from the land. An important exception in the Indian Ocean is the so-called Dipole, which describes variability in the difference between western and eastern equatorial Indian Ocean sea-surface temperatures.[16] Although the two sides of the ocean vary independently of each other, the name Indian Ocean Dipole has become standard. The Dipole, and the tropical Indian Ocean more generally, have important effects on climate over parts of Australia and East and Southern Africa, and play an important role in how El Niño affects some areas beyond the Pacific.[17]

8.2.1.1.4 Extratropical oceans

In the tropics, sea-surface temperature anomalies have an impact on the climate primarily because of large changes in evaporation; in the extratropics the impact on changes in air temperature and air pressure may be more important than on evaporation. The changes in air temperature and air pressure can affect storm tracks, intensities and frequencies, because of effects of the temperature and pressure gradients on the jet streams (§ 4.2.8.2).[8] Similar effects occur at the sea-ice boundary: the retreat of sea-ice through warming may therefore have important effects on climate in the mid- and high-latitudes.[18]

8.2.1.2 The land

While changes in tropical sea-surface temperatures are the main reason why forecasters can predict the general weather conditions over the next few months, land and ice/snow conditions should not be completely ignored (§ 5.2.7). For example, after an unusually dry period, the land surface may dry up, and in the summer months it can be heated to unusually high temperatures (a larger proportion of the sun's energy is used to increase the temperature rather than to heat and evaporate soil, plant and surface water). The resultant hot dry air provides a basis for predicting heat waves in places such as Europe[19,20] and South Asia.[21] However, a dry land-surface is usually insufficient in itself to provide a strong basis for seasonal forecasts. Instead it may act to reinforce effects of sea-surface temperature anomalies, as was the case in the US Dust Bowl of the 1930s,[22] for example.

8.2.1.3 Snow and ice

Like land temperature and soil moisture, ice and snow cover have some predictable influences on the weather at seasonal timescales. You may need to wear sunglasses after snow because it acts like a mirror, reflecting the sunlight into your face and back into space. Because this sunlight is reflected rather than absorbed, there is less heating of the surface than if there were no snow or ice, and so the overlying air is not heated much. This cooling can result in more snow or freezing, even more sunlight is then reflected, and further cooling occurs. This albedo effect (§ 4.2.7) is important at multi-year timescales (see Chapter 9), but less so in seasonal forecasting, partly because the year-to-year differences are rather small and only a small proportion of the Earth is covered in ice and snow. However, year-to-year differences can be useful in predicting spring snow-melt in places such as California. Similarly, unusually heavy snowfall over the Himalayas in winter can slow the summer heating of inland South Asia, and thus weaken the summer monsoon (See Box 7.3).

8.2.2 How are seasonal forecasts made?

The principle behind seasonal forecasting is to predict how unusual conditions at Earth's surface (detailed in § 8.2.1) might affect the persistence, frequency and/or intensity of certain weather types (such as rainstorms). The effects of Earth's surface on the seasonal climate are most likely to be detectable if: a) the influence on the overlying air is strong (as is likely in the tropics given large sea-surface temperature anomalies); and b) if this *boundary forcing* persists for a long period. The target period (Box 7.1) must therefore not be too short lest individual weather events, which are unpredictable, mask the effects of the boundary forcing; but the target period must not be too long, lest those surface conditions change unpredictably. Because of these constraints, in practice, seasonal forecasts are rarely made for periods of less than two or three months, and generally for not much longer than four or five months.

The effects of boundary forcing on the climate are modelled using one of two approaches, but in both cases the starting point is with observations of Earth's surface. In contrast, the starting point for weather forecasting is with observations of the initial conditions (§ 7.4). One of the approaches to seasonal forecasting – *empirical modelling* – effectively addresses the question of how similar boundary conditions in the past have affected climate. The other approach – *dynamical modelling* – considers how, in principle, the current boundary forcings might affect the climate.

8.2.2.1 Empirical prediction

Some early, and pre-scientific, methods of seasonal forecasting were based on observations that some climate anomalies seemed to be pre-figured by other climate anomalies or unusual occurrences in nature. Who has not asked questions such as whether we can anticipate an unusually hot summer given the dry winter or abundant spring blossoms we may have just experienced, or whether the coming wet season will be delayed given the cold winds of the last few weeks? Unfortunately, few of these types of observations provide any robust basis for forecasting, and they are often highly subjective.

With our improved understanding of how climate operates, empirical relationships between anomalous surface conditions and subsequent climate, when supported by theoretical considerations, can be used with confidence as a basis for making forecasts. The commonest approach is to use some form of regression or classification procedure to relate observations of sea-surface temperature anomalies, including those associated with El Niño and La Niña (Box 5.1), with climate anomalies over the following few months. A simple example, might consider how climate has been affected by episodes of El Niño conditions in the past. Depending on the sophistication of the statistical model used, sea-surface temperatures in areas beyond the equatorial Pacific may also be considered, as well as observations of other boundary forcings (§ 8.2.1). In fact, the first such model was developed in 1886 to forecast the Indian monsoon, and was based purely on observations of Himalayan snowfall.[6] However, it took almost 100 years before empirical models became more widely adopted. They are used extensively today by many countries to make their national seasonal forecasts, most of which are based on sea-surface temperatures in various parts of the tropical oceans.

8.2.2.2 *Dynamical prediction*

The second approach to seasonal forecasting is similar to the way in which weather forecasts are made (Chapter 7). Just as weather forecasting relies heavily on numerical models, so dynamical methods of seasonal forecasting use climate models. In many respects, a climate model is virtually identical to a numerical weather prediction model (NWP; see Box 8.2), and so the forecasting procedure follows the same steps as for weather prediction: observation, analysis, initialization, integration and post-processing. The details of each of these steps depend on the complexity of how Earth's surface is represented in the climate model (Box 8.2).

BOX 8.2 CLIMATE MODELS

A climate model is very similar to an NWP model (§ 7.4). However, there are some differences between the two types of models; the most important one for the purposes of seasonal forecasts is that a climate model needs to have a reasonably realistic representation of conditions at Earth's surface. In particular, some representation of the ocean is required, although the most sophisticated models also include ice and land-surface components. Given the importance of sea-surface temperatures in seasonal forecasting (§ 8.2.1) only the representation of the oceans is discussed here.

There are different ways of representing the oceans in a climate model, ranging in complexity from simply specifying the surface conditions, to modelling the ocean in a similar way to modelling the atmosphere. A key distinction is whether or not the surface conditions are predicted independently of making the seasonal climate forecast, or whether the two are predicted together. If the surface conditions are predicted first a two-tiered forecast system is used; if the surface conditions are predicted as part of the climate forecasting step then a one-tiered system is used.

Two-tiered systems

In a two-tiered forecast system the sea-surface temperatures are predicted first, and then these predicted temperatures are prescribed when making the seasonal forecast.[23] The sea temperatures may be predicted as simply as by persisting the current anomalies over the next few months, or gradually damping them towards average.[24] Despite their simplicity, it is difficult to produce forecasts that are a lot more accurate than such procedures for about the first three months. For predicting further than three months, more sophisticated forecasts of sea-surface temperatures are required. These forecasts can be made either by empirical procedures or by running dynamical models of the oceans. One problem with two-tiered forecasting is that the oceans are allowed to affect the atmosphere, but changes in wind and temperature, etc., are not able to affect the oceans adequately (§ 8.2.1.1). Some climate models incorporate very simple models of the oceans, perhaps allowing changes in temperature and evaporation, but not having ocean currents. However, if seasonal forecasts are to consider the development of phenomena such as El Niño and La Niña then a proper ocean model is required.

One-tiered systems

One tiered forecasting systems use 'coupled models',[25] i.e., a model for predicting the atmosphere is run together with a model for the ocean. The ocean models need to be run at higher spatial resolution than the atmosphere because of the importance of small-scale features in the circulation of the oceans. This requirement, together with the importance of initializing the ocean model (§ 8.2.2), means that forecasting with coupled models requires some of the most powerful computers in the world. Only a few of the Global Producing Centres (§ 8.3.1) are able to run such models.

8.2.2.2.1 Observation

With a seasonal forecast, the critical problem is not so much to get the initial atmospheric conditions correct (since those will be lost after the first week or two of the forecast), but instead to get the *boundary conditions* and their effects correct. To a reasonable approximation the boundary conditions are a mathematical representation of the boundary forcings described above. Observations of Earth's surface are therefore required, and most notably of sea-surface temperatures.

With the advent of the satellite era, estimates of sea-surface temperatures over the global oceans have become available. These estimates do not have the same problems as do the satellite estimates of land-surface temperatures (§ 6.3.2.2) except in the presence of persistent cloud. To calibrate and supplement the satellite measurements, arrays of moored buoys have been implemented in the most important areas of the oceans. The Tropical Atmosphere Ocean (TAO) / Triangle Trans Ocean Buoy Network (TRITON) array is in the Pacific Ocean,[26] and was motivated by a failure in 1982 to recognize that the largest El Niño then on record was developing. Similar arrays have been implemented in the tropical Atlantic[27] and Indian[28] Oceans. The moored arrays are supplemented by a set of drifting buoys with other automated instruments.

The various instruments measure more than just sea-surface temperatures; they also take weather observations (winds, atmospheric pressure, etc.; § 6.3.1.1), and measure ocean currents and salinity down to 500 m or more beneath the surface. These additional measurements are important for initializing ocean models

(Box 8.2), and provide essential information if seasonal forecasts beyond about three months are required. Because of their remoteness, maintenance of the moored buoys is difficult and so data gaps can last weeks if they stop transmitting. The buoys are frequently targets of vandalism and theft, especially those near South America, The resulting data gaps can create problems for forecasting and monitoring.

Observations of sea-ice and land-surface conditions (primarily soil moisture and snow cover) have received far less attention for seasonal forecasting than have observations of the oceans. Soil moisture is still poorly measured and is rather estimated from rainfall and soil properties. Snow and ice can be measured by satellite, although it is much easier to measure extent than thickness. Because of the poor availability of data, and the relatively weak influence of soil moisture and snow and ice, such observations are used by only a few centres in operational seasonal forecasting.

8.2.2.2.2 Analysis

The analysis from weather forecasting can be used for seasonal forecasts, but an ocean analysis is also required if an ocean model is being used. The method for generating an ocean analysis is similar to that for the weather analysis, although the procedure is a little more complicated for the ocean, in part because of poorer data availability and because of the greater inertia in the oceans (winds can change much more easily than ocean currents). Because of the greater uncertainty in the state of the ocean compared to the atmosphere, multiple analyses are made; in contrast, a single atmospheric analysis is typically generated. These ocean analyses are then perturbed to produce an ensemble (Box 7.6).

8.2.2.2.3 Initialization

Regardless of the complexity of the climate model's representation of Earth's surface (Box 8.2), an ensemble approach (Box 7.6) is essential when forecasting seasonal climate using dynamical models. For weather forecasting, the generation of perturbations in the initial conditions is a critical step because predicting the evolution of the current weather is the basis for skilful weather forecasts. For seasonal forecasting, however, there is little pretence that the current weather conditions provide much useful information for predicting weeks and months into the future, and so simpler ensemble generation methods such as lagged averaging (Box 7.6) are widely used. In fact, until only a few years ago, the majority of forecasting centres did not attempt to initialize their models with recent observations of the weather at all.[24] However, initializing a climate model using recent weather observations does become more important if a dynamical ocean model is being used (Box 8.2), or if a seamless prediction system is being operated.[29] In a seamless prediction system, weather, sub-seasonal and seasonal forecasts may all be made in one process.

8.2.2.2.4 Integration

Seasonal forecasts are much more uncertain than weather forecasts, and if representing uncertainty is important in weather forecasting, it is even more so at seasonal timescales. In weather forecasting, the main sources of uncertainty are the initial conditions and the model errors. For seasonal forecasting, the model errors remain as a source of uncertainty, while the initial conditions are effectively forgotten after about two weeks, but an additional source of uncertainty is that Earth's surface has only a limited effect on how the weather varies. As for weather forecasting, these sources of uncertainty can be addressed, at least in part, by appropriate use of ensembles (§ 7.4.4).

The uncertainty from the boundary forcing is addressed by using a large ensemble, but also by considering the forecast over a period of a few months rather than a shorter period of weeks. Thus, if the surface conditions are conducive to the formation of strong and frequent cyclones, for example, strong and frequent cyclones may be detectable because of the large sample size, even if the effect of the surface is weak. However, because of imperfections in climate models, the response to the boundary forcings may not be simulated adequately, and so a large sample of predictions from one model may not be helpful, and may even be misleading (spurious effects of the surface on climate could be simulated). A multi-model ensemble is therefore recommended. Multi-modelling is less important for weather forecasting, but has a clear advantage over a large single-model ensemble at seasonal and longer timescales.[30]

Because seasonal forecasting requires the models to be run much further into the future than for weather forecasting, climate models are typically run at a coarser resolution than for an NWP model. Running at high resolution can be prohibitively expensive. In numerical weather prediction, regional models are used widely to provide more detailed, and hopefully more accurate, forecasts at national or regional scale. Such downscaling models have not been widely used for seasonal forecasts, partly because of computational expense, and partly because of a lack of clearly demonstrated additional benefit over simpler, empirically-based downscaling methods[31] (see further discussion in the Post-processing section below).

8.2.2.2.5 Post-processing

As implied in the previous discussion on model integration, one problem when running a climate model for many weeks is that there are differences between the climate of the model and the climate in the real world because of imperfections in the model. The predictions tend to drift fairly quickly towards the model's climate and away from more realistic conditions. This problem of drift has been particularly severe for climate models that are coupled to dynamical ocean models, although major improvements have been made in the last few years.[32]

The simplest way to correct for problems of model drift is to express the forecast with reference to the model's own (lead-time dependent) climate. For example, the forecast may be presented as an anomaly compared to the model's climatological average (as is widely performed for forecasts of ENSO), or probabilistic forecasts may be derived with reference to the model's climatological terciles (Box 8.3). These procedures are widely adopted for global seasonal forecasts, but do not guarantee good reliability[29] (see Box 7.4 for a technical definition of reliability). As a result, the forecast probabilities from many forecasting centres cannot be taken at face value, and have to be interpreted carefully using detailed diagnostics of model skill (see further discussion in § 8.4). This task of interpretation is difficult even for experts.

BOX 8.3 TERCILE FORECASTS

Because of the large uncertainties inherent in making seasonal forecasts, they are generally presented as probabilistic rather than as deterministic forecasts (Box 7.5). The most common predictands (Box 7.1) are three- or four-month rainfall accumulations, and average temperatures. The forecast indicates the probabilities that the accumulation or average for the predicted period will fall within pre-defined ranges. These pre-defined ranges, or categories, are derived by considering historical values from a recent 30-year climatological period (§ 4.3).

The most common practice is to define three categories. If, the rainfall is more than the 10th wettest within a 30-year climatology, that rainfall would be classified as 'above-normal', whereas if it is less than the 10th driest it would be classified as 'below-normal'. If the rainfall is neither more than the tenth wettest, nor less than the tenth driest then it is classified as 'normal'. An example is shown in Table 8.2, using the same rainfall data from Case Study 5.1, i.e., December–February rainfall over Botswana. (For simplicity, December 1980–February 1981 is listed as 1981.) The tenth wettest year was 1994, with 300 mm. This threshold approximates the 'upper tercile'; one-third (i.e., ten) of the years had 300 mm or more. (There are various ways of calculating the terciles, and a more exact value is somewhere between that for 1981 and 1994.) Similarly, the tenth driest year was 2003 with 200 mm. This threshold approximates the lower tercile; one-third of the years had 200 mm or less. Following this standard, there are equal numbers of years in each category.

Typically, the range of the 'normal' category is narrow, and so 'above-normal' and 'below-normal' may not be particularly extreme (see an additional example in Figure 4.5, where the 'normal' category is bounded by the two vertical dotted lines). In most cases, it is reasonable to interpret 'normal' as 'close-to-average', although if the data are strongly positively skewed (as may be the case for rainfall in arid areas) it is possible for rainfall to be 'above-normal' and still be below-average.

TABLE 8.2 Country-averaged December–February rainfall accumulations for Botswana

Ordered by year			*Ordered by rainfall*		
Year	*Rainfall*	*Category*	*Year*	*Rainfall*	*Category*
1981	326	A	1992	153	B
1982	162	B	2002	158	B
1983	163	B	1982	162	B
1984	174	B	1983	163	B
1985	222	N	2007	169	B
1986	214	N	1987	170	B
1987	170	B	1995	171	B
1988	330	A	1984	174	B
1989	354	A	1998	197	B
1990	219	N	2003	200	B
1991	268	N	2001	208	N
1992	153	B	2005	213	N
1993	223	N	1986	214	N
1994	300	A	1990	219	N
1995	171	B	1985	222	N
1996	342	A	2004	223	N
1997	320	A	1993	223	N
1998	197	B	1999	240	N
1999	240	N	2010	267	N
2000	439	A	1991	268	N
2001	208	N	1994	300	A
2002	158	B	1997	320	A
2003	200	B	2008	321	A
2004	223	N	2009	325	A
2005	213	N	1981	326	A
2006	416	A	1988	330	A
2007	169	B	1996	342	A
2008	321	A	1989	354	A
2009	325	A	2006	416	A
2010	267	N	2000	439	A

Data source: *Climate Prediction Center Merged Analysis of Precipitation [CMAP]*

An alternative way of addressing the problem of drift, and of model systematic errors more generally, is to apply some form of model output statistics (MOS) correction, similar to that applied in the post-processing step of numerical weather prediction (§ 7.4.5). Applying an MOS correction is effectively a hybrid approach combining empirical and dynamical prediction, and it can act as a *downscaling* method (i.e., a method of generating a more detailed forecast) in the same way as for weather forecasting. The main difficulty is sample size: seasonal forecasts are typically updated about once per month instead of four times per day for weather forecasts. Forecasters can pretend to turn back the clock and make 'forecasts' for periods that are now in the past (*hindcasts*). However, even generating a reasonable number of hindcasts is difficult because of computational expense, but more so because of the unavailability of important observational data prior to both the installation of oceanic arrays (see the section on Observation) and the satellite era. Without these important observations the models cannot be initialized well. However, sample sizes of past forecasts are steadily growing, and post-processing schemes are becoming increasingly popular options for regional and national forecasting.[33] If calibrated properly, the seasonal forecast probabilities are more reliable than model outputs that are adjusted only for errors in climatology. Unfortunately, this gain in reliability makes the forecasts look weak much of the time[34] (although appropriately so).

Sample size problems are also an issue when combining predictions from multiple models. In theory one would expect that a better forecast could be obtained if the more skilful models were given greater consideration than the weaker ones. In practice, however, it is difficult to demonstrate that one model is unequivocally better than another given only a small number of forecasts. It is therefore hard to improve upon treating each model equally, although perhaps after selecting a subset of what appear to be the better models.

8.3 What seasonal forecasts are available?

Since the late-1990s, the World Meteorological Organization (WMO) has been facilitating and directing the establishment of a seasonal forecasting infrastructure to support countries around the world to make regular operational forecasts. The vision for this infrastructure is similar to that of the World Weather Watch (§ 6.2), and is now being implemented through the Global Framework for Climate Services.[35] The main components of this infrastructure are described in the following sub-sections.

8.3.1 Global Producing Centres of Long-Range Forecasts

In 2006, the WMO established a process for designating centres that produce global seasonal forecasts and that make these products available to countries for producing their own official forecasts. There are currently 13 of these Global Producing Centres of Long-Range Forecasts (GPCs),[36,ii] most of which provide a range of publicly available forecast products. Their predictions are collected by the Lead-Centre for Multi-Model Ensembling,[iii] which produces multi-model predictions that are accessible by National Meteorological and Hydrological Services, and a few graphical products that are accessible publicly. These forecasts are updated monthly. As part of the designation process, a set of hindcasts has to be produced and verification information must be provided to the Lead-Centre for Standardized Verification System of Long-Range Forecasts.[iv] Most of this verification information is likely to be too technical for most purposes in public health work.

Two additional centres provide global seasonal forecasts, but are not formally designated as GPCs: the Asia-Pacific Economic Cooperation (APEC) Climate Center (APCC), and the International Research Institute for Climate and Society (IRI). The APCC has played an important role in research on multi-model forecasting. It has recently been developing some experimental sub-seasonal forecasting products (Box 7.2) that are available publicly. The IRI has a long history of producing multi-model seasonal forecasts, and is one of only a few centres that calibrates its forecasts in an effort to provide reliable forecast probabilities (Box 7.4). As well as a forecasting centre, the IRI is a World Health Organization Collaborating Center (USA 430) for early warning systems for malaria and other climate sensitive diseases.[v] As such, the IRI engages in developing climate services for particular health issues – such as Zika virus transmission in the Americas (see Case Study 8.1).

CASE STUDY 8.1 UNDERSTANDING AND PREDICTING LATIN AEDES-BORNE DISEASES IN LATIN AMERICA AND THE CARIBBEAN USING CLIMATE INFORMATION

Madeleine C. Thomson and Ángel G. Muñoz, IRI, Columbia University, New York, USA

The Zika virus (ZIKV), which emerged in Brazil in 2015 to devastating effects, is principally transmitted globally, and in Latin America and the Caribbean (LAC), by the container breeder mosquito *Aedes aegypti*. *Aedes albopictus* (the Tiger mosquito) is identified as a possible significant future vector because of its recent rapid spread.[37] Both species also transmit dengue fever, chikungunya and yellow fever viruses, and other viral diseases making their presence in the region a significant public health concern.

Ae. aegypti is common in urban environments in the tropics and sub-tropics. Its success comes from its preference for ovipositing in both natural and artificial water-filled receptacles, where the eggs can survive when the water contents fluctuate and regularly expose them to drying conditions.[38] Although ZIKV transmission depends on several factors including human behaviour, temperature is a significant driver of the development rates of juvenile *Aedes aegypti* and *Aedes albopictus* and adult feeding/egg-laying cycles along with the length of extrinsic incubation period and viral replication of arboviruses.[39] Both drought and excess rainfall have been implicated in the creation of indoor and outdoor breeding sites for *Aedes* vectors of ZIKV and associated epidemics of dengue and chikungunya. Climate-based early warning systems for dengue, a related virus that is transmitted by the same vectors, have been suggested in different regions of the world.

The ZIKV epidemic that emerged in Brazil in 2015 occurred during a period of exceptionally high temperatures and drought in association with an El

Niño event. However, the extreme climate anomalies observed in most parts of South America during the 2015 epidemic were not caused exclusively by El Niño or climate change, but by a combination of climate signals acting at multiple timescales.[40]

Aedes vectors can respond to both unusually dry and wet conditions (as they may switch from indoor domestic breeding sites to outdoor sites made temporarily available). This change in dominant breeding sites has implications for the development of control measures for wet or dry years. Given the importance of climate's year-to-year variability in determining ZIKV risk potential, an early warning prototype for environmental suitability of *Aedes spp.* disease transmission was developed using a basic reproduction number -R0- model,[41] combined with state-of-the-art seasonal forecasts from the North American Multi-Model Ensemble (NMME).[42] Using this approach, probabilistic predictions of above-normal, normal or below-normal potential risk of transmission for at least the following season (three months) are made. Such information is deemed useful for health practitioners and other decision-makers. The predictive capacity is highest for multiple countries in LAC during the December–February and March–May seasons and it is slightly lower – although still of potential use to decision-makers – for the rest of the year. It is important to emphasize that although it is possible at seasonal timescale to forecast suitable environmental conditions for transmission of *Aedes*-borne diseases, this does not mean that the actual transmission, or even epidemic events, are predictable with this type of forecast system.

8.3.2 Regional Climate Centres and Regional Climate Outlook Forums

Regional Climate Outlook Forums (RCOFs) were initiated in 1997 to promote the production of an authoritative seasonal forecast through regional consensus.[43] They have since been established in most regions of the world,[44,45] and are an important source of seasonal forecasts. The Forums themselves are intended as a means of interaction between forecast producers and users,[46] and may serve as an opportunity for public health specialists to provide feedback on existing forecast products and their broader needs for climate services. In Southern Africa, the Malaria Outlook Forum (MALOF) was initiated by the health community after evidence emerged of the significant predictability of malaria incidence using climate information.[47] Instead of the malaria control managers attending the RCOF, members of the climate community participated in the preparatory malaria meeting, which was held immediately prior to the rainy season. At this meeting the seasonal climate forecast was integrated directly into the planning process for the coming malaria season.[48] The MALOF was run successfully between 2004 and 2007, but was then discontinued when external donor funding dried up.

While also subject to fluctuating donor funds, the RCOFs have become a routine fixture in the annual climate services calendar. They are convened up to three times per year in a given region, but monthly updates to the seasonal forecasts are coordinated by Regional Climate Centres,[vi] and participating National Meteorological and Hydrological Services (NMHSs) are encouraged to update their own national forecasts.

There is broad diversity amongst the RCOFs, but all produce a standard seasonal forecast indicating probabilities of 'below-normal', 'normal' and 'above-normal' accumulated rainfall and/or average temperature (§ 4.3.3 and Box 8.3). The probabilistic tercile-based formats have been justly criticized as being unnecessarily obtuse and of minimal relevance,[44] but are likely to remain the staple output of the RCOFs for the foreseeable future. These forecasts may be of some value for indicating *suitable* climate conditions for supporting pathogens, pests and diseases (§ 7.2), but are unlikely to be good indicators of *hazardous* or *inhospitable* conditions. For example, even perfect seasonal forecasts of rainfall would still be poor indicators of flood risk because of a weak relationship between seasonally accumulated rainfall and flooding.[49] A few of the RCOFs are attempting to address these limitations by developing experimental forecasts of extreme events.[50] However, as discussed in § 8.4 the reliability (Box 7.4) of many of the RCOF forecasts is problematic, and rigorous skill assessments have yet to be published. Some of the RCOFs make archives of past forecasts available so that some assessment of their value in specific applications can be conducted.

8.3.3 National meteorological and hydrological services

In general, climate services in most countries are much less well-developed than weather services. This lack of capacity is being addressed by the Global Framework for Climate Services,[35] but the capacity of many NMHSs to interact with public health specialists is likely to be limited, especially in developing countries. Nevertheless, most countries where there is some predictability of seasonal climate (see § 8.4) have developed a seasonal forecasting capability through their participation in the RCOFs.

The WMO is promoting the production of national climate watches, along similar lines to weather alerts (§ 7.6.1). These watches will act as official national alerts of developing and expected hazardous climate conditions. As with the weather alerts, it is important to work with alerts from mandated government authorities. Standards for climate watches have yet to be set, but some countries have implemented systems already. The watches combine monitoring and forecast information, and are particularly useful for slow-onset hazards like drought.

8.3.4 Additional global products

The Global Seasonal Climate Update[vii] is intended as a quarterly forecast and monitoring product that is coordinated by the WMO. The Update is not yet fully operational, and is initially targeted at Regional Climate Centres and NMHSs to assist in the production of their own information products. However, the intention is for the Update to be of value to public users with global and regional-scale interests.

The idea for the Global Seasonal Climate Update emerged from interest in WMO's El Niño/La Niña Updates. These Updates summarize the current status of the ENSO (see Box 5.1) and review the available predictions. Amongst other sources, the El Niño/La Niña Update draws from the Climate Prediction Center/ IRI joint ENSO Diagnostic Discussion. This information is updated monthly, and provides a comprehensive review of ENSO forecasts from around the world.

8.4 Do seasonal forecasts work well?

A key message when considering the possible value of seasonal forecasts is that, unlike weather forecasts, seasonal forecasts only work in some parts of the globe, for some times of the year, and even for some years. Whereas a weather forecast should be available every day, seasonal forecasts may only provide indications of possible anomalous climate conditions occasionally. The reason for this intermittency is simple: in many places and for much of the time the climate is insufficiently affected by unusual conditions at Earth's surface. Consider ENSO, for example, which is the primary reason seasonal forecasts are possible (§ 8.2.1). Despite its importance, ENSO affects less than a third of global land areas in a predictable way,[51] and El Niño and La Niña episodes occur less than half the time. Therefore, just like volcanoes, ENSO (and other influences on the climate) provides a basis for making a seasonal forecast only sometimes. In addition, the frequency and intensity of El Niño and La Niña vary inter-decadally and inter-millennially,[52,53] and the predictability of ENSO, and of climate variability in general, is relatively poor during the quiescent phases. We have been experiencing a relatively active phase of ENSO since the last few decades of the 20th century, but in the mid-20th century, ENSO variability was relatively weak, and some early operational seasonal forecasts performed poorly in this period. It is unclear how long this current active period, and therefore this period of good seasonal predictability, will last.

If seasonal forecasts do not work everywhere, where do they work? An indication is provided in Figure 8.1, which illustrates the skill (Box 7.4) of IRI's seasonal forecasts over the last 20 years. Since the forecasts are in the standard probabilistic format (Box 8.3), measuring the skill in a simple manner is a non-trivial matter (Box 7.4). Figure 8.1 uses a measure that scores the forecasts well if the observed category had a high probability rather than only considering whether the category with the highest probability occurred. Hence, if the observed category had a 60% probability that forecast will score more highly than if the forecast probability was 50%.

The skill of the seasonal temperature forecasts (top) is notably higher than that for rainfall (bottom). (Note that the grey-scales are different for the two maps; the

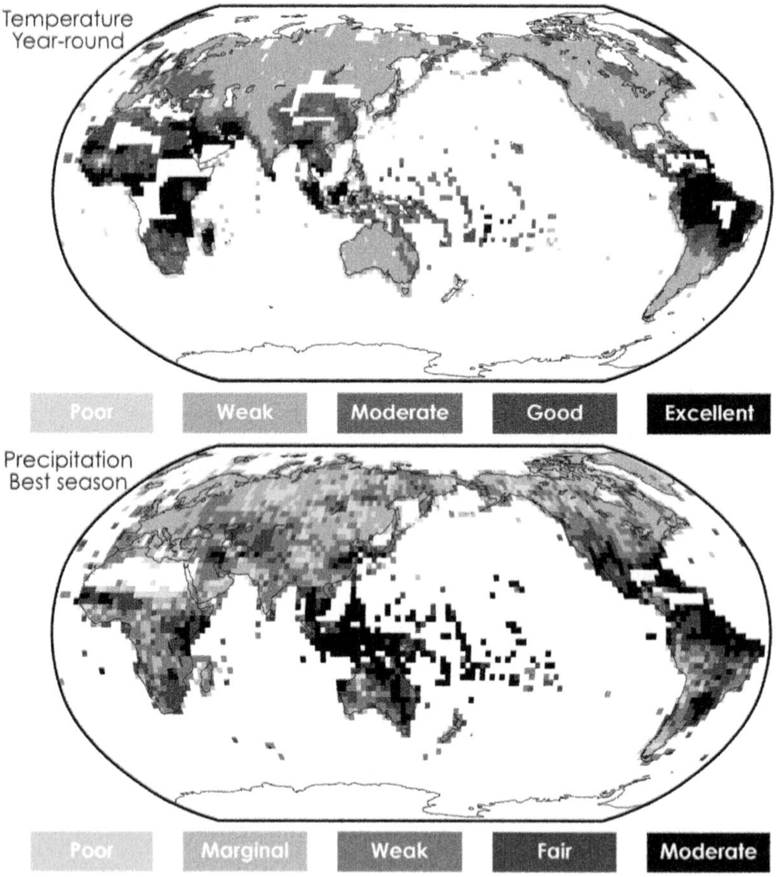

FIGURE 8.1 A measure of value of IRI's seasonal (three-month) average temperature (top) and accumulated rainfall (bottom) forecasts for 1997–2017. The value is estimated by calculating the percentage return on investments on IRI's shortest lead-time probabilistic forecasts if paid out with fair odds.[54] For temperature, the value is calculated using forecasts year-round, and the score is classified as 'excellent' (> 30%), 'good' (20–30%), 'moderate' (10–20%), 'weak' (1–10%) or 'poor' (< 1%). For rainfall, only the value for the highest-scoring season is shown, and the score is classified as 'moderate' (> 10%), 'fair' (5–10%), 'weak' (2.5–5%), 'marginal' (1–2.5%) or 'poor' (< 1%). Climatological forecasts would score 0%.

online colour versions of the maps use consistent colours and so the difference in skill is easier to see.) This difference is actually under-represented by the maps since the skill for the temperature forecasts is calculated using forecasts throughout the year, whereas for rainfall the skill is shown only for the season that is easiest to forecast. One reason why the temperature forecasts are much better than the seasonal forecasts is because of global warming. Above-normal temperatures have occurred far more frequently than below-normal over at least the last 20 years as a result of global warming, and because the seasonal forecasts have been able to predict this ongoing trend, they have been scored well. The usefulness of such information is open to debate, and discussions about how best to communicate variability in temperature from year-to-year rather than the longer-term trends would be beneficial.

The skill of the rainfall forecasts is generally poor partly because of the difficulty of predicting accumulated rainfall. At the weather forecasting timescale, it is harder to predict rainfall amount than occurrence (§ 7.5.2); this difficulty translates into the seasonal timescale. A simple principle is that it is easy to forecast (at all timescales) a meteorological parameter that is spatially coherent than one that is localized. Since rainfall intensity tends to be highly localized, whereas rainfall occurrence is more coherent, intensity is harder to predict. For the same reason, it may be possible to make more accurate forecasts of climate impacts, such as crop yields[55] or disease incidence,[47] directly from the drivers of climate variability rather than using the predicted climate per se. Perhaps regrettably, seasonal accumulations, rather than some other measure, continue to be the main predictand for seasonal rainfall forecasts. Some of the RCOFs are beginning to experiment with new seasonal forecast products such as numbers of wet-days and wetspells,[46] but products of this nature are not yet part of standard practice.

One other strong message from Figure 8.1 is that, in general, seasonal forecasts work much better in the tropics than in the extratropics.[56] This pattern is the opposite of that for weather forecasts (Figure 7.1); weather forecasts work better in the extratropics than in the tropics. The better quality of the seasonal forecasts in the tropics is a result of the stronger effect of sea-surface temperatures there (§ 8.2.1).

The IRI's forecasts are carefully calibrated, and so the forecast probabilities are broadly reliable.[34] However, unfortunately, the same cannot be said of some of the RCOF and national forecasts,[57] where there has been an ongoing problem in assigning too much probability to the 'normal' category. That is not to say that these forecasts are unskilful, but only that care needs to be taken in interpreting the probabilities.

The health community has a strong focus on using evidence-based policies and practices. An evidence-based approach should also be taken when incorporating climate information into health decision-making. With increased knowledge about the way forecasts are made, their strengths and their limitations, we believe that the health community will be better placed to work with climate scientists to improve the transparency, relevance and quality of the information provided.

8.5 Conclusions

Seasonal climate forecasts are predictions of the general rather than the specific weather conditions of the coming few months. In contrast to weather forecasts, seasonal forecasts generally work better in the tropics than in the extratropics, but even in the tropics, seasonal forecasts are only useful intermittently, and there are many parts of the world where there is no skill at all. Most available operational seasonal forecasts may be of some value for indicating suitable climate conditions for supporting pathogens, pests and diseases, but further research is required to assess their value as indicators of hazardous or inhospitable weather conditions over the coming few months. Despite their limitations, seasonal forecasts create a gap in forecast skill between the next few days and the next few months. This gap is being explored through research on sub-seasonal forecasting. There is another gap between seasonal forecasts and long-term climate change projections; timescales beyond seasonal are discussed in the subsequent chapter.

Notes

i https://www.nytimes.com/interactive/projects/spot-the-ball/2014/06/17.
ii A map is available online: www.wmo.int/pages/prog/wcp/wcasp/gpc/gpc.php.
iii https://www.wmolc.org/.
iv www.bom.gov.au/wmo/lrfvs/.
v http://apps.who.int/whocc/.
vi www.wmo.int/pages/prog/wcp/wcasp/rcc/rcc.php.
vii https://www.wmo.int/pages/prog/wcp/ccl/opace/opace3/documents/GSCU-Brief.pdf.

References

1 Inwards, R. *Weather Lore: A Collection of Proverbs, Sayings and Rules Concerning the Weather.* 190 (Senate, London, 1994).
2 Mason, S. J. in *Seasonal Forecasts, Climatic Change and Human Health, Advances in Global Change Research (AGLO)* (eds. M. C. Thomson, R. Garcia-Herrera, & M. Beniston), 13–29 (Springer, New York, 2008).
3 Vecchi, G. A. *et al.* Statistical–dynamical predictions of seasonal North Atlantic hurricane activity. *Monthly Weather Review* **139**, 1070–1082 (2011).
4 Weisheimer, A. & Palmer, T. N. On the reliability of seasonal climate forecasts. *Journal of The Royal Society Interface* **11**, 20131162 (2014).
5 Lowe, R. *et al.* Evaluating probabilistic dengue risk forecasts from a prototype early warning system for Brazil. *Elife* **24**, doi:10.7554/eLife.11285 (2016).
6 Slingo, J. & Palmer, T. Uncertainty in weather and climate prediction. *Philosophical Transactions of the Royal Society A* **369**, 4751–4767 (2011).
7 Goddard, L. *et al.* Current approaches to seasonal to interannual climate predictions. *International Journal of Climatology* **21**, 1111–1152 (2001).
8 Palmer, T. N. & Anderson, D. L. The prospects for seasonal forecasting – a review paper. *Quarterly Journal of the Royal Meteorological Society* **120**, 755–793 (1994).
9 Namias, J., Yuan, X. & Cayan, D. R. Persistence of North Pacific sea surface temperature and atmospheric flow patterns. *Journal of Climate* **1**, 682–703 (1988).
10 Kushnir, Y. Interdecadal variations in North Atlantic sea surface temperature and associated atmospheric conditions. *Journal of Climate* **7**, 141–157 (1994).
11 Cane, M. A., Zebiak, S. E. & Dolan, S. C. Experimental forecasts of El Niño. *Nature* **321**, 827–832 (1986).
12 Zebiak, S. E. & Cane, M. A. A Model El-Niño Southern Oscillation. *Monthly Weather Review* **115**, 2262–2278 (1987).
13 Vaughan, L. F., Furlow, J., Higgins, W., Nierenberg, C. & Pulwarty, R. US Investments in international climate research and applications: reflections on contributions to interdisciplinary climate science and services, development, and adaptation. *Earth Perspectives Transdisciplinarity Enabled* **1**, 23, doi:10.1186/2194-6434-1-23 (2014).
14 Wang, C. Atlantic climate variability and its associated atmospheric circulation cells. *Journal of Climate* **15**, 1516–1536 (2002).
15 Ward, M. N. Diagnosis and short-lead time prediction of summer rainfall in tropical North Africa at interannual and multi-decadal timescales. *Journal of Climate* **11**, 3167–3191 (1998).

16 Saji, N. H., Goswami, B. N., Vinayachandran, P. N. & Yamagata, T. A dipole mode in the tropical Indian Ocean. *Nature Letter* **401**, 360–363, doi:10.1038/43854 (1999).
17 Goddard, L. & Graham, N. E. The importance of the Indian Ocean for simulating precipitation anomalies over Eastern and Southern Africa. *Journal of Geophysical Research* **104**, 19099–19116 (1999).
18 Cohen, J. *et al.* Recent Arctic amplification and extreme mid-latitude weather. *Nature Geoscience* **7**, 627 (2014).
19 van den Hurk, B. *et al.* Soil moisture effects on seasonal temperature and precipitation forecast scores in Europe. *Climate Dynamics* **38**, 349–362 (2012).
20 Fischer, E. M., Seneviratne, S. I., Vidale, P. L., Lüthi, D. & Schär, C. Soil moisture–atmosphere interactions during the 2003 European summer heat wave. *Journal of Climate* **20**, 5081–5099 (2007).
21 Nissan, H., Burkart, K., Mason, S. J., Coughlan de Perez, E. & van Aalst, M. Defining and predicting heat waves in Bangladesh. *Journal of Applied Meteorology and Climatology* **56**, 2653–2670, doi:10.1175/JAMC-D-17-0035.1 (2017).
22 Schubert, S. D., Suarez, M. J., Pegion, P. J., Koster, R. D. & Bacmeister, J. T. On the cause of the 1930s Dust Bowl. *Science* **303**, 1855–1859 (2004).
23 Barnston, A. G., Kumar, A., Goddard, L. & Hoerling, M. P. Improving seasonal prediction practices through attribution of climate variability. *Bulletin of the American Meteorological Society* **86**, 59–72 (2005).
24 Mason, S. J. *et al.* The IRI seasonal climate prediction system and the 1997/1998 El Niño event. *Bulletin of American Meteorological Society* **80**, 1853–1874 (1999).
25 Stockdale, T. N., Anderson, D. L. T., Alves, J. O. S. & Balmaseda, M. A. Global seasonal rainfall forecasts using a coupled ocean-atmosphere model. *Nature* **392**, 370–373 (1998).
26 McPhaden, M. J. *et al.* The Tropical Ocean – Global Atmosphere Observing System: a decade of progress. *Journal of Geophysical Research* **103**, 14169–14240 (1998).
27 Bourlès, B. *et al.* The PIRATA program: history, accomplishments, and future directions. *Bulletin of the American Meteorological Society* **89**, 1111–1125 (2008).
28 McPhaden, M. J. *et al.* RAMA: the research moored array for African–Asian–Australian monsoon analysis and prediction. *Bulletin of the American Meteorological Society* **90**, 459–480 (2009).
29 Vitart, F. *et al.* The new VAREPS-monthly forecasting system: a first step towards seamless prediction. *Quarterly Journal of the Royal Meteorological Society* **134**, 1789–1799 (2008).
30 Hagedorn, R., Doblas-Reyes, J. F. & Palmer, T. N. The rationale behind the success of multi-model ensembles in seasonal forecasting. Part I: basic concepts. *Tellus A* **57**, 219–233 (2005).
31 Robertson, A. W., Qian, J. H., Tippett, M. K., Moron, V. & Lucero, A. Downscaling of seasonal rainfall over the Philippines: dynamical versus statistical approaches. *Monthly Weather Review* **140**, 1204–1218 (2012).
32 Reichler, T. & Kim, J. How well do coupled models simulate today's climate? *Bulletin of the American Meteorological Society* **89**, 303–311 (2008).
33 Manzanas, R., Lucero, A., Weisheimer, A. & Gutiérrez, J. M. Can bias correction and statistical downscaling methods improve the skill of seasonal precipitation forecasts? *Climate Dynamics* **50**, 1161–1176 (2018).
34 Barnston, A. G. *et al.* Verification of the first 11 years of IRI's seasonal climate forecasts. *Journal of Applied Meteorology and Climatology* **49**, 493–520 (2010).
35 Hewitt, C., Mason, S. & Walland, D. The global framework for climate services. *Nature Climate Change* **2**, 831–832 (2012).
36 Graham, R. J., Yun, W.T., Kim, J., Kumar, A., Jones, D., Bettio, L., Gagnon, N., Kolli, R.K. & Smith, D. Long-range forecasting and the Global Framework for Climate Services. *Climate Research* **47**(1/2), 47–55 (2011).
37 Messina, J. P. *et al.* Mapping global environmental suitability for Zika virus. *eLife* **5**, e15272, doi:10.7554/eLife.15272 (2016).

38 Faull, K. J. & Williams, C. R. Intraspecific variation in desiccation survival time of Aedes aegypti (L.) mosquito eggs of Australian origin. *Journal of Vector Ecology* **40**, 292–300, doi:10.1111/jvec.12167 (2015).
39 Brady, O. J. *et al.* Global temperature constraints on Aedes aegypti and Ae. albopictus persistence and competence for dengue virus transmission. *Parasites & Vectors* **47**, 388, doi:10.1186/1756-3305-7-338 (2014).
40 Muñoz, A. G., Thomson, M. C., Goddard, L. & Aldighieri, S. Analyzing climate variations at multiple timescales can guide Zika virus response measures. *GigaScience* **20165**, 41, doi:10.1186/s13742-016-0146-1 (2016).
41 Caminade, C. *et al.* Global risk model for vector-borne transmission of Zika virus reveals the role of El Niño 2015. *Proceedings of the National Academy of Sciences* **114**, 119–124, doi:10.1073/pnas.1614303114 (2017).
42 Kirtman, B. P. *et al.* The North American multimodel ensemble: phase-1 seasonal-to-interannual prediction; phase-2 toward developing intraseasonal prediction. *Bulletin of the American Meteorological Society* **95**, 585–601 (2014).
43 Buizer, J. L., Foster, J. & Lund, D. Global impacts and regional actions: preparing for the 1997–98 El Nino. *Bulletin of the American Meteorological Society* **81**, 2121–2139 (2000).
44 Patt, A. G., Ogallo, L. J. & Hellmuth, M. Learning from 10 years of climate outlook forums in Africa. *Science* **318**, 49–50 (2007).
45 Ogallo, L., Bessemoulin, P., Ceron, J.-P., Mason, S. & Connor, S. J. Adapting to climate variability and change: the Climate Outlook Forum process. *Bulletin WMO* **57**, 93–102 (2008).
46 Gerlak, A. K. *et al.* Building a framework for process-oriented evaluation of Regional Climate Outlook Forums. *Weather, Climate, and Society* **10**, 3–4 (2018).
47 Thomson, M. C. *et al.* Malaria early warnings based on seasonal climate forecasts from multi-model ensembles. *Nature* **439**, 576–579 (2006).
48 Thomson, M. *et al.* Climate and health in Africa. Special Issue: The International Research Institute for Climate & Society: Shaping the Landscape of Climate Services. Section 1: IRI Approach & Examples of Integrated Research & Demonstration. *Earth Perspectives: Transdisciplinarity Enabled* 1, 17 (2014).
49 Coughlan de Perez, E. *et al.* Should seasonal rainfall forecasts be used for flood preparedness? *Hydrology and Earth System Sciences* **21**, 4517 (2017).
50 Hamilton, E. *et al.* Forecasting the number of extreme daily events on seasonal timescales. *Journal of Geophysical Research: Atmospheres* **117** (2012), doi.org/10.1029/2011JD016541.
51 Mason, S. J. & Goddard, L. Probabilistic precipitation anomalies associated with ENSO. *Bulletin of the American Meteorological Society* **82**, 619–638 (2001).
52 Diaz, H. F. & Markgraf, V. E. *El Niño: Historical and Paleoclimatic Aspects of the Southern Oscillation* (Cambridge University Press, Cambridge, 1992).
53 Enfield, D. B. & Cid S, L. Low-frequency changes in El Nino-southern oscillation. *Journal of Climate* **4**, 1137–1146 (1991).
54 Hagedorn, R. & Smith, L. A. Communicating the value of probabilistic forecasts with weather roulette. *Meteorological Applications* **16**(2), 143–155 (2009).
55 Cane, M., Eshel, G. & Buckland, R. Forecasting Zimbabwean maize yield using eastern equatorial Pacific sea surface temperature. *Nature* **16**, 3059–3071 (1994).
56 Kim, G. *et al.* Global and regional skill of the seasonal predictions by WMO Lead Centre for Long-Range Forecast Multi-Model Ensemble. *International Journal of Climatology* **36**, 1657–1675 (2016).
57 Mason, S. J. & Chidzambwa, S. *Verification of RCOF Forecasts.* Report No. 09-02, 23pp (International Research Institute for Climate and Society, Palisades, New York, 2008).

9

CLIMATE INFORMATION FOR ADAPTATION

From years to decades

Hannah Nissan, Madeleine C. Thomson, Simon J. Mason and Ángel G. Muñoz
Contributors: Glynn Vale, John W. Hargrove, Arthur M. Greene and Bradfield Lyon

> The blessed gods
> Purge all infection from our air while you
> Do climate here!
>
> The Winter's Tale *by William Shakespeare*

9.1 Introduction

Climate change was likened to the fifth horseman of the apocalypse by Margaret Chan in November 2007 in the first speech on the subject by a World Health Organization (WHO) Director. The following May, health protection from climate change was identified as the priority issue of concern at the 2008 World Health Assembly through a resolution ratified by its 193 member states.[1] The threat that a changing climate poses to global health was starkly elaborated by the 2009 Lancet Commission on 'Managing the Health Effects of Climate Change'.[2] While direct effects of climate change on the health of vulnerable populations, e.g., from heat waves, were a major concern the indirect impact of climate change on water, food security and extreme climate events was considered to pose a greater health threat.

Managing the effects of climate change on human health will require inputs from all sectors of government and civil society, as well as collaboration between many academic, governmental, private sector and community organizations. Effective management must be informed by the best available science and, in part, this requires the use of long-term projections of the climate over the coming century. Long-term projections are particularly important for the development of policies to *mitigate* climate change through reductions in carbon emissions. These reductions have measurable health co-benefits. Long-term projections can help quantify the extent to which these co-benefits can compensate for the mitigation cost of achieving the targets of the Paris climate agreement.[3]

Weather and climate vary on multiple timescales (§§ 3.2 and §§ 5.3), and climate information (historical, current or future) must target the specific time and space scales of the decisions being made. Observed climate is the result of the interaction of natural climate variability and the anthropogenic climate-change signal associated with increasing greenhouse gas emissions (§ 5.4.2). Today's climate is dominated by natural variability, but the climate-change signal is already emerging and is expected to strengthen as concentrations of greenhouse gases in the atmosphere increase. However, gradual, long-term trends in climate are not the means by which people will experience most aspects of climate change. Instead, impacts will be felt primarily through changes in the weather (including extreme events like heat and cold waves and extreme rainfall), the seasons and potentially through alterations to longer-term components of climate variability, such as the El Niño – Southern Oscillation (see Box 5.1). The predictability of all these different timescales varies by location and period under consideration (see Chapters 7 and 8).

In writing this book, we have prioritized shorter timescales of weather and climate variability because these have most direct relevance to the types of operational and planning decisions made in the health sector. However, in this chapter, we focus on longer-term health decisions that require climate knowledge and information at timescales beyond a year to multiple decades (up to 50 years). This time span incorporates both natural decadal variability and the influences of anthropogenic climate change. It is an extremely challenging timeframe at which to work because of the very limited operational predictive skill at multi-annual timescales, and limited confidence about how rainfall may change on longer timescales. Despite these challenges, the timescales under consideration, especially the next five to 20 years, are particularly important when considering how to *adapt* to climate change. Here we discuss what we do and do not know about the climate in the coming years. We then focus on the practical use of climate information in understanding observed climate impacts on health as well as the prediction of multi-annual to multi-decadal risks. The multi-annual timescale is often referred to as 'decadal', even though the forecasts are for periods less than a decade.

9.2 How increasing concentrations of CO_2 can impact health

Instead of presenting a comprehensive overview of the causes and consequences of anthropogenic climate change (see Box 9.1 for a brief history of the science), we illustrate some of the ways by which health is affected by focusing on one of the main greenhouse gases of concern, namely CO_2. The steady increase of CO_2 in our atmosphere, associated with burning fossil fuels, deforestation and agricultural practices, is routinely observed.[4] Its impact on the environmental determinants of health manifests through a number of pathways (Figure 9.1). We now explore how CO_2 in the atmosphere and oceans ultimately impact the three health concerns we have focused on throughout this book, namely the health impacts of disasters, infectious diseases and nutrition.

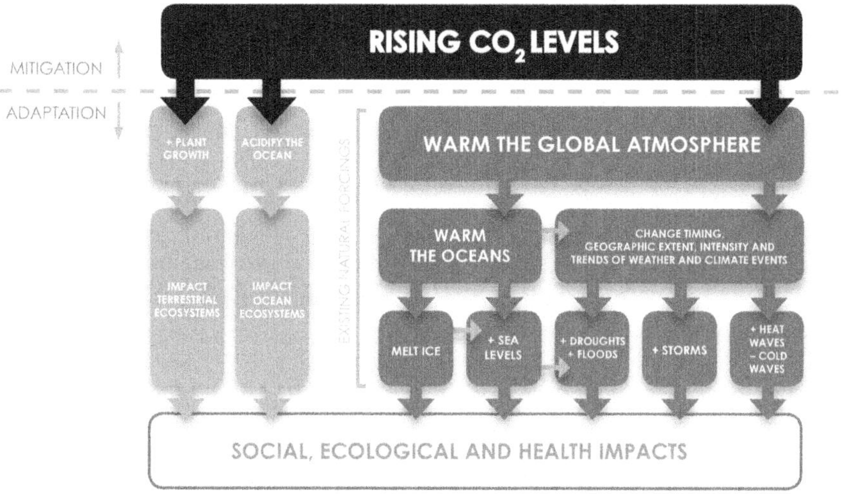

FIGURE 9.1 Impact pathways of rising CO_2 on social, ecological and health outcomes. Pathways include plant fertilization, ocean acidification and a warming atmosphere. The latter has a direct and indirect (via oceanic warming) impact on our climate. Both atmospheric and ocean warming impact on sea levels through thermal expansion of the seas, destabilization of coastal land ice, and melting of land ice and snow.

BOX 9.1 A BRIEF HISTORY OF THE SCIENCE OF GLOBAL WARMING

The origins of the science of global warming

The physical basis for what we now call climate change was established in the 19th century. As early as 1861, the physicist John Tyndall[5] provided empirical evidence of the critical role of greenhouse gases (including CO_2) in maintaining Earth's temperature.[6] His findings demonstrated the importance of CO_2 and water vapour in trapping heat in Earth's climate system (see Chapter 4). He established the physical basis for the first prediction of the magnitude of expected global warming as a result of increasing CO_2 levels, made by Svante Arrhenius in 1896.[7]

Thus, climate-change science is based on the physical laws of properties of gases in the air – laws which have been known for well over 100 years. Evidence of these expected changes are increasingly being observed through extensive analyses not only of climate, but also of environmental and impact data from around the world.[8]

Mauna Loa carbon dioxide monitoring

In 1958, Charles Keeling began collecting data on CO_2 in the air at the Mauna Loa Observatory, in Hawaii, and on Antarctica.[9] The Antarctica site was discontinued because of lack of funding, but the site at Mauna Loa has been operating continuously to this day. The data from Mauna Loa constitute the oldest continuous record of atmospheric CO_2. This record is known as the Keeling Curve, and indicates an increase in CO_2 concentrations from 315 ppmv in 1958 to 407 ppmv in 2017. Keeling's data provided the first significant evidence that concentrations of CO_2 in the air were increasing. Subsequent measurements of CO_2 trapped in air bubbles in ice cores indicate that CO_2 concentrations prior to the industrial era were around 275–285 ppmv. The increase in CO_2 since the pre-industrial era is therefore about 45%.

Intergovernmental Panel on Climate Change (IPCC)

In 1988, the Intergovernmental Panel on Climate Change (IPCC) was formed 'to provide policymakers with regular assessments of the scientific basis of climate change, its impacts and future risks, and options for adaptation and mitigation'.[i] The primary outputs of the IPCC are Assessment Reports, in which available literature and published analyses documenting the state of climate change and its impacts are reviewed. These Assessment Reports are updated approximately every five to seven years. Additional special reports and supporting documents provide further information. Since its First Assessment Report in 1990, the IPCC has involved thousands of scientists around the world in pushing forward the frontiers of climate science, estimating the economic costs and benefits of mitigation and, in the Fourth Assessment and subsequent reports, identifying needs and opportunities for adaptation.

9.2.1 Hydro-meteorological disasters

During 26–28 August 2017, Hurricane Harvey poured over a trillion gallons of water over Texas, causing unprecedented floods. The rainfall directly impacted millions of people whose lives and livelihoods were put at risk. The capacity of the weather services to predict the development and movement of the storm (see § 7.5.3), thus forewarning the population and emergency response teams, came about because of massive investments in computational capacity for weather modelling in Europe and the USA. The disastrous 2017 hurricane season for the Caribbean and USA raises a critical question: are tropical cyclones becoming more extreme and is this a consequence of climate change?

It remains unclear whether observed changes in tropical cyclone activity have exceeded natural variability to date, in part because of strong decadal variability in activity, at least in the North Atlantic.[10] Tropical cyclones are driven by energy from the warmth of the sea-surface (see § 4.2.8). Evidence from modelling and theory indicates that tropical cyclones will become stronger, larger and more destructive in the future as the oceans warm in response to increasing atmospheric temperatures.[11] However, other factors are involved in hurricane and typhoon occurrence, and there is substantial variation among projections of how the frequency of such storms may change in the future. Modelling studies suggest that the global average frequency of tropical cyclones may decrease, but the most intense cyclones could become more frequent.[12] For example, although higher concentrations of CO_2 are expected to increase peak hurricane intensity during future La Niña years in the Atlantic, changes of wind patterns in the same ocean could suppress hurricane activity during El Niño events.[13]

9.2.2 Infectious diseases

Trends in average climate (particularly temperature) as well as changes in extreme weather events and seasonality resulting from increasing greenhouse gas concentrations have already been detected. However, because of the paucity of both historical climate and health data over decadal and longer timescales (i.e., > 30 years), evidence showing how observed climate-change trends have influenced disease transmission at a local level is rare. The scarcity of historical health data from developing countries has meant that a few datasets are used repeatedly in a large number of studies. These include malaria data from Kericho, Kenya[14–18] and cholera data from the International Center for Diarrhoeal Disease Research (ICDDR), Bangladesh.[19–21] A detailed database has been developed for Zimbabwe, in which daily meteorological information is matched with 60 years of data on disease vectors (see Case Study 9.1).

CASE STUDY 9.1 TSETSE – CHANGES IN CLIMATE IN THE ZAMBEZI VALLEY: IMPACT ON TSETSE FLIES

John W. Hargrove and Glynn Vale, South African Centre for Epidemiological Modelling and Analysis (SACEMA), University of Stellenbosch, South Africa

Rekomitjie Research Station (16°8′S, 29°25′E, altitude 503 m), in the Zambezi Valley of Zimbabwe, was founded in 1959 for studies of the ecology and behaviour of tsetse flies, which are the vectors of the trypanosomes that cause the diseases of sleeping sickness in humans and nagana in livestock. The location was chosen because the two species of tsetse present, namely *Glossina pallidipes* and *G. morsitans morsitans*, could be caught in numbers large enough to facilitate investigations.[22] Moreover, since the station is in the Mana Pools National Park, which was designated a United Nations Educational,

Scientific and Cultural Organization (UNESCO) World Heritage Site in 1984, the area has been largely free of anthropogenic changes to woodland cover and tsetse hosts, leading to the expectation that large catches of tsetse would be maintained indefinitely. However, in the 1990s the annual average catches began to show a net downward trend,[23] and in the last five years the abundances of *G, pallidipes* and *G. m. morsitans* have been the lowest on record, at only 1% and 5%, respectively, of the levels recorded in the 1960s (Vale, pers. comm.). Climate change is suspected of being the main cause of this population decline.

Daily meteorological records taken at the station show no material change in rainfall in the last 57 years, but annual average minimum and maximum temperatures have increased by 1.81 °C and 1.08 °C, respectively (Figure 9.2).[24] Most of the rise in minimum temperature has occurred in the dry months of May to November. Changes in maximum temperature have taken place during the hot dry season of September to November (Figure 9.2), when the maximum temperature is at least 32 °C on most days and can occasionally exceed 40 °C. The greatest increase in temperatures was in November, at the end of the hot dry season, when daily minima and maxima rose by 2.62 °C and 2.44 °C, respectively. The increase in temperatures during the hottest months was not due simply to an increase in the temperature of unusually hot days, with maxima up to 44 °C, but due also to a greater frequency of such days. An appreciation of the details of these temperature variations is essential for understanding the impact of climate change on the dynamics of tsetse populations. The hot dry season is commonly a period in which tsetse abundance declines by around 60–80%, with reductions of > 90% if the season is especially hot,[23,25] due largely to the direct physiological effects of high temperature, which raise the death rate of the flies much more than they enhance the birth rate.[26] The problem is especially severe with the immature flies, i.e., pupae and very young adults. Moreover, high temperatures are associated with increased rates of parasitism and predation of pupae.[26] The upshot is that breeding becomes progressively less successful as daily maximum temperatures rise above 35 °C. As a result, at the end of the hot dry season there is a marked paucity of young flies in the adult population.[26] The reduction in tsetse abundance during the hot dry season is potentially significant since tsetse can breed only slowly,[26] even in the favourable weather of cooler months, so limiting the ability of population numbers to recover after a knock-back.

The salient point is that the greatest increases in temperature have occurred at the end of the hot dry season – the very time when tsetse populations are most vulnerable to a greater intensity and duration of heat stress.

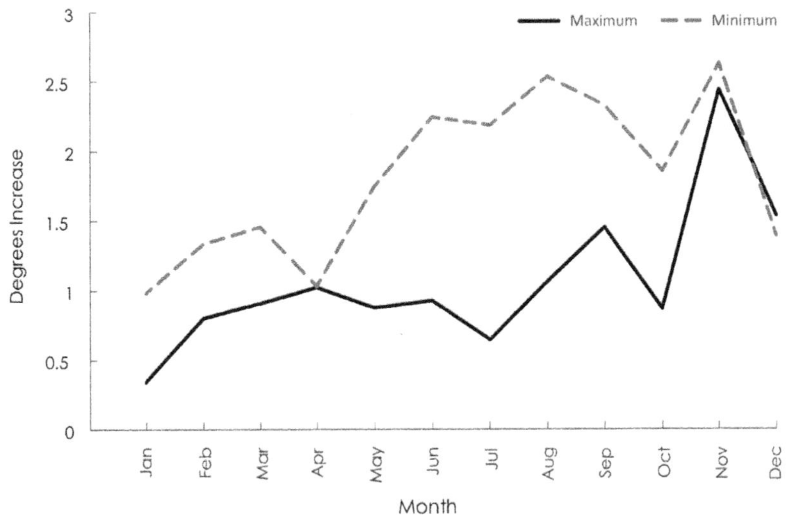

FIGURE 9.2 Increase in the average minimum and maximum temperatures of each calendar month, from 1960 to 2016, at Rekomitjie Research Station in Zimbabwe. Data from linear regression where the dependent variable was either the mean maximum or mean minimum, in the pertinent month of each year, and the independent variable was the year since 1960. The increases in the minima are significant ($P < 0.05$) for all months except April. For the maxima the increases are significant only for August, September and November.

Observational studies that use less than 30 years of data risk confusing long-term climate-change trends with natural decadal variations over ten to 30 year timescales (Box 9.2). The challenge that decadal climate variability poses to forecasting future long-term climate risks are elaborated in § 9.4.2. An example of the practical impact of decadal variability on decision-making is outlined in Case Study 9.2.

9.2.3 Nutrition

CO_2 is an important trace gas of the atmosphere and part of the natural carbon-cycle. It is the sole source of carbon for photosynthesis, by which it is converted to carbohydrates by plants, which are then use as food or fibre by all manner of creatures, including ourselves. Under normal conditions most plants are carbon-hungry and will readily convert additional atmospheric CO_2 to plant growth. The gas is often added to greenhouses to increase yields; a process called the 'fertilization

BOX 9.2 FILTERING THE CLIMATE SIGNAL BY TIMESCALES

Arthur M. Greene and Ángel G. Muñoz IRI, Columbia University, New York, USA

It may be useful, for planning or adaptation purposes, to understand the way in which the different components of observed climate variability combine to produce the resulting 'signal' that we experience. For example, if, in a particular region, rainfall variations tend to be dominated by year-to-year (*interannual*) variations while decadal-scale variability is relatively weak, attention can be focused on prediction of, and adaptation to, interannual swings. On the other hand, the presence of strong decadal variations may prompt research into attribution of its causes in an attempt to shed light on potential future variations, as well as adaptation measures that might be undertaken with those longer time-horizons in mind.

In order to facilitate our understanding of how observed variations may be thus decomposed by timescale, a Maproom was created in 2010,[27] accessible via the Data Library of the International Research Institute for Climate and Society (IRI). The decomposition process technically consists of a linear regression (a projection) of the original rainfall or temperature time series onto an anthropogenic climate-change signal, followed by successive filtering operations.[27,28] The resultant series comprises three components: a) a climate-change signal; b) decadal, or low-frequency variation; and c) interannual fluctuations. The first of these may be thought of as a 'drift' term, describing long-term variations linked to the anthropogenic climate-change signal. When this first component is subtracted from the original time series, the result is a new series that contains the rest of the timescales. This secondary series is then filtered using a particular window (size of the filter), yielding the 'decadal' component. Subtracting this component in turn yields the high-frequency, or interannual, component of the original series.

This approach was used during the 2014–2016 Zika epidemic to contextualize the role of climate. Contrary to what was being assumed at the time, a combination of signals involving El Niño, climate change and other climate drivers were responsible for setting suitable conditions for the transmission of Zika, with implications for disease response measures.[29]

effect'. One might argue, as some do, that there are positive food-security benefits from human-induced increases in atmospheric CO_2. While enhanced crop yields for C3 plants, such as rice, wheat, barley and soya bean, are a potentially positive outcome of increased atmospheric CO_2, the actual impact is highly dependent on changes in temperature and rainfall that will occur alongside CO_2 increases.

CASE STUDY 9.2 THE EAST AFRICAN PARADOX

Bradfield Lyon, University of Maine, USA, Madeleine C. Thomson, IRI, Columbia University, New York, USA

Over roughly the past two decades East Africa has experienced an increasing frequency of drought, particularly during the 'long rains' season, which typically runs from March to May. This increasing frequency of drought is linked to an overall downward trend in East African rainfall that has been underway since the 1980s. Some climate scientists have argued that this drying trend is associated with an upward trend in sea-surface temperature, especially in the tropical Indian Ocean.[30] In simplest terms, the argument is that higher ocean temperatures lead to increased rainfall over the ocean that ultimately robs East Africa of its moisture and rain. Meanwhile, climate-change projections suggest that the climate of East Africa will become wetter, not drier, by the end of the current century.[31]

This situation presents an apparent contradiction, with observations over recent decades indicating more frequent drought conditions while long-term climate projections suggest that the region will in fact become wetter. This 'East Africa Paradox' highlights two important factors for assessing climate-change impacts: considering climate variability on multiple timescales, and understanding the strengths and limitations of climate-change projections. Rather than a gradual decline in rainfall since the 1980s, the long rains of East Africa have instead undergone an abrupt decline that occurred around 1998–1999.[32] At the same time, similarly abrupt changes in sea-surface temperatures, mainly in the tropical Pacific Ocean, were observed (Figure 9.3). Such a decadal shift does not preclude a longer-term wetting trend as indicated by the IPCC,[31] but highlights the importance of distinguishing between the two timescales, as adaptation plans developed for a wetter future could leave the region more vulnerable to drought in the shorter term.

The inability of coupled models to capture key physical drivers of East African climate, particularly conditions in the tropical Pacific and Indian oceans, undermines confidence in the future climate projections of the region. Anthropogenic climate change could thus potentially contribute to drying rather than wetting. When these issues were presented to the executive board of Roll Back Malaria in a workshop in Tanzania in 2014, the group of experts recommended that attention be paid by the malaria community to the risks associated with a return of higher rains in the region in the medium term. In addition to the possible public health impacts, such a change could impact perceptions of the success of malaria programs and pose a risk to donor confidence.

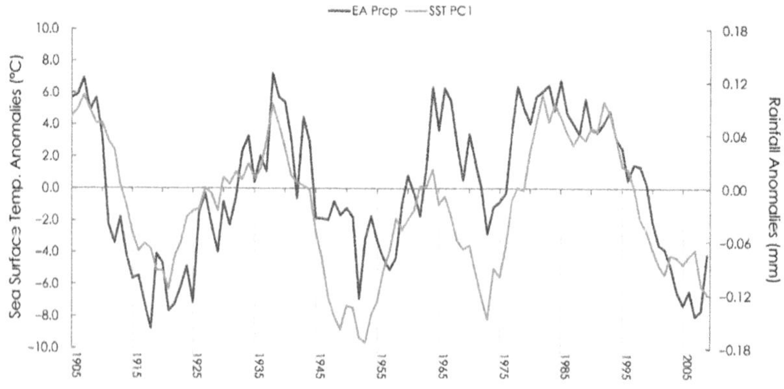

FIGURE 9.3 Sea-surface temperature anomalies and March–May (MAM) rainfall anomalies in Eastern Africa. Time series of the leading sea-surface temperature mode (anomalies in °C, see left scale) and the detrended East African rainfall anomaly (mm/month, right scale) for MAM. A nine-year moving average has been applied to both series (1905–2008). Adapted from Lyon (2014) and extended in time

Besides, crops are not the only plants to be affected: there is a fertilization effect of CO_2 on poison ivy photosynthesis, and a shift toward a more allergenic form of urushiol.[33] The fertilization effect may also have negative consequences for human nutrition as it has been associated with a 5–10% reduction in micronutrients such as iron and zinc and lower protein content of food staples.[34]

The warming from the enhanced greenhouse effect has a direct impact on the climate system and is the primary cause of sea-level rise (Figure 9.1). Global sea levels have been increasing at the rate of about 15 mm per decade (primarily measured using tide stations and satellite laser altimeters). Because CO_2 remains in the atmosphere for a long time, this rise is expected to continue in the coming centuries even if global emissions are seriously curtailed. For now, the predominant cause of sea-level rise is due to the thermal expansion of the water in the oceans and the melting of ice sheets and glaciers on land. More than 10% of the world's population live in low-elevation coastal zones prone to floods from sea-level rise.[35] Many of these regions are already experiencing the impacts of climate change and are trying to prepare for worse to come. In particular, the impact of sea-level rise on Small Island Developing States (SIDS), which risk being inundated, has long been recognized.[36] Coastal flooding results in increased salinity of drinking water, decreased habitable and agricultural land area, disrupted fisheries and diminished food security.

Sea-level rise is not the only negative impact of increasing CO_2 on the sea: when CO_2 dissolves in seawater it makes the oceans more acidic. This decline in oceanic pH has a damaging effect on marine ecosystems and the food chains that rely on them. Ocean warming and acidification have the potential to impact the quality and quantity of seafood with follow-on effects for future food security and ecosystem stability.[37]

9.3 How climate-change projections are made

The starting point for generating climate-change projections is to set plausible socioeconomic story-lines for the coming century based on economic and population forecasts.[38] These trajectories are then translated into emissions scenarios with associated concentrations of greenhouse gases and aerosols. Finally, the effects on the global climate are modelled using general circulation models (GCMs; similar to those used for seasonal forecasting [§ 8.2.2]), which produce projections of future climate using the greenhouse gas and aerosol concentrations specified by these scenarios.[39] Probabilistic projections are generated for each scenario by running simulations using multiple models with different initial conditions (see Box 7.6).

The models used to predict weather and climate from days to decades ahead are similar, but the complexity of physical processes and the components of the Earth system that must be included in the models increases with lead-time. For example, seasonal forecast models must have some representation of the ocean as it is the sea-surface temperatures that are most important for predictions on this timescale, whereas some numerical weather prediction models do not consider changing surface conditions (see Table 5.1 and § 5.4). The models used for climate-change projections must include full ocean models, since the circulation of the deep ocean is a crucial component of longer-term climate change. Just as for the models used in both weather and seasonal forecasting, the models used for climate-change projections involve simplifications (*parameterizations* – § 7.4.4). Because of the need for these parameterizations, features such as clouds, radiation processes, carbon chemistry and small-scale weather features, such as thunderstorms, local rainfall and flooding are not explicitly simulated, and so global climate models should not be used to directly infer information about local climate.[40]

9.3.1 Downscaling

Information is often required at smaller spatial and temporal scales than can be provided by global models, to assist with local, national or regional decision-making. Output from GCMs can be 'downscaled' dynamically (using a limited area regional climate model, RCM) or statistically (using statistical relationships determined from historical observations). Dynamical high resolution RCMs (such as the PRECIS model developed by the UK Met Office, or the WRF model developed by the National Center for Atmospheric Research) take the large-scale climate fields provided by the global models as boundary conditions and use the same fundamental laws of physics to simulate the climate on a grid of (approximately) 1–50 km^2. Regional models are commonly used to develop projections of the future climate at the national level. In statistical downscaling, which is less resource intensive, quantitative relationships are developed between local climate variables (e.g., near-surface air temperature and rainfall) and large-scale predictors (such as pressure fields). These statistical functions are then applied to the output of GCMs to simulate the future climate at a higher resolution. The post-processing steps of weather and seasonal forecasting are equivalent to statistical downscaling (§§ 7.4.5 and 8.2.2). The two approaches have different advantages and disadvantages (see Table 9.1).

TABLE 9.1 The main advantages and disadvantages of downscaling using statistical methods or regional climate models (RCMs)

Statistical downscaling	*Dynamical downscaling (RCMs)*
Main advantages: • Requires fewer computational resources • A wide variety of methods can be used • Bias correction is an integral part of the process	**Main advantages:** • Within the RCM domain, individual variables and inter-variable dependencies are physically consistent in time and space • The same fundamental physical laws are used in both an RCM and a GCM • RCMs do not assume stationarity in the climate system except for sub-grid parameterizations • No specific calibration data are required (though evaluation with observations is essential)
Main disadvantages: • Requires a long meteorological record • Any quality problems in the calibration data will be transferred to the downscaled data ('rubbish in, rubbish out') • The climate is changing, but this approach assumes that statistical relationships between large-scale predictors and local climate remain stationary • The higher the requirements regarding spatial, temporal and inter-variable consistency, the more complex and computationally demanding the statistical procedures become	**Main disadvantages:** • RCMs require substantial computational resources • RCMs have their own biases and errors, which compound the problems inherited from the parent GCM • Near the boundary of the RCM domain, spurious effects can occur

9.3.2 Multi-annual to multi-decadal prediction

It is often surprising to learn that climate experts are more confident in their predictions of the 30-year average climate towards the end of the century than the evolution of the climate over the next ten years. However, a similarly counter-intuitive increase in confidence with longer lead-times was noted when comparing seasonal forecasts with long-term weather forecasts (§ 8.2). Recall from Chapter 7 that the weather is unpredictable beyond about seven to ten days because of errors in the initial conditions. Nevertheless, the climate (defined here as the statistics of weather over a period of time; Box 4.1) can be predicted with some skill at a range of lead-times (see § 8.2). For longer-term climate prediction, it is the current state of the land surface and oceans, rather than more volatile atmospheric conditions, that provide a basis for prediction.

Multi-annual to multi-decadal prediction (two to thirty years ahead), is at the forefront of climate research because of the value of this timeframe for adaptation planning and the inherent forecasting challenges involved.[41] On this timescale, both natural interannual-to-decadal variability and long-term trends are important. Near-term climate change is a transition timeframe: these lead-times combine the difficulties of seasonal and long-term climate-change projections, as both initial conditions (primarily in the oceans and land surface) and externally forced trends play a role (see Table 9.2).

The Coupled Model Intercomparison Project (CMIP)[42] is a key activity in research on multi-annual to multi-decadal predictions. The Project started in 1995 as a means to compare climate models. It has emerged as a powerful resource to advance model development and scientific understanding of the Earth system through systematic comparisons of climate model outputs from multiple climate modelling centres. To meet its current objectives CMIP has developed well-defined protocols for climate model simulations. The most recent set of protocols, CMIP5, sets out to promote a standard set of model simulations in order to:

- evaluate how well the models reproduce the recent past.
- make available model projections of future climate change on two timescales, *near-term* (out to about 2035) and *long-term* (out to 2100 and beyond), and.
- improve understanding of differences in model projections.

The long-term CMIP5 (Coupled Model Intercomparison Project Phase 5) projections have been a key input to the WHO's Climate and Health Country Profile reports (Box 9.3).

There are important differences between the CMIP5 long-term projections developed for the IPCC, and the near-term predictions used in the experimental development of multi-annual forecasts (currently limited to two to nine-year lead-times; see Table 9.2). The near-term predictions require initialized runs (see §§ 7.4.3 and 8.2.2 on initialization) that start from current observations of the climate and predict how the system will evolve from there. The long-term projections

BOX 9.3 CLIMATE AND HEALTH COUNTRY PROFILES

The WHO has published a series of Climate and Health Country Profile reports[ii] which provide country-specific estimates of current and future climate hazards and the expected burden of climate change on human health. Country-specific, national level, time series plots of specific projections are provided up to the year 2100 for climate hazards using CMIP5 projections. A high-emissions 'business as usual' Representative Concentration Pathway 8.5 (RCP8.5) scenario is compared to projections under a 'two-degree' scenario with rapidly decreasing emissions following Representative Concentration Pathway 2.6 (RCP2.6). The profiles include several plots of future climate conditions: mean annual surface temperature, warm spell days, days with extreme rainfall and consecutive dry days. For some countries, cold spell days and warm nights are also presented. In addition, the profiles track current national policy responses and identify opportunities for health co-benefits from climate mitigation actions. While these figures do give some representation of uncertainty it is impossible to estimate how accurate these forecasts may be.

TABLE 9.2 Projections of near-term climate change (multi-annual to multi-decadal) with CMIP5

	Multi- decadal projections	*Multi-annual predictions*
Model simulations used	Uninitialized CMIP5 model projections for the IPCC	Interannual-to-decadal forecasts from the Decadal Climate Prediction Project initialized with current observations
Processes simulated	Anthropogenically-forced trend; natural decadal variability is simulated but its timing does not align with the real world	Anthropogenically-forced trend; natural decadal variability
Model run time-line	Centennial lead-times	2- to 9-year lead-times
Modelling approach	Multi-model ensemble (from CMIP5) for a variety of anthropogenic emissions scenarios	Multi-model ensemble (subset of CMIP5 models) initialized using current observations
What the models can tell you	Long-term trends and the statistics of climate over three or more decades, but cannot be used in forecasting individual decades	Trend and decadal variability; where skilful, they could be used in forecasting

are unitialized. These runs simulate natural variability, but; without initialization, these cycles are not in phase with the real world, so they cannot be used to forecast climate impacts at specific dates. Instead, they can be used to infer statistics about the climate over several decades (30 years or more). In both cases, a multi-model ensemble average (§ 7.6) is often calculated in an attempt to iron out model errors. However, in the case of the uninitialized projections, this average smoothes out the natural decadal variability simulated by each model, and therefore only captures the climate-change trend. The natural variability around this trend is an additional source of uncertainty in these projections and must be factored in if they are to be used effectively in decision-making.

An approach to incorporating climate change into long-term malaria planning that considers these uncertainties is illustrated here:

> *Climate-proofing the malaria eradication strategy should begin with an assessment of the vulnerabilities of existing plans to climate variability and change. Identifying and reducing these vulnerabilities will ensure that plans are robust to uncertainty by avoiding the more common approach of tailoring decisions to specific (and highly uncertain) projections of future climate change. Analyses of climate variability and change in specific areas and on relevant timescales can then be conducted* [Box 9.1]. *Confidence in future climate projections can only be assessed through thorough model evaluation, focused on understanding the timescales and locations for which models perform well and, conversely, where they fail. To place trust in future projections, it further needs to be demonstrated that when models do succeed, they do so for the right reasons, by capturing the appropriate physical processes and large-scale drivers of local impacts. Such assessments require bespoke analyses targeting specific regions and applications, and cannot be shortcut through a one-size-fits-all approach.*[43]

The example of the East Africa Paradox (Case Study 9.2) illustrates the pitfalls of an off-the-shelf method.

9.4 How accurate are multi-annual to multi-decadal forecasts?

9.4.1 Climate model errors

Climate models have been evaluated extensively; they can reproduce many of the most important aspects of the climate. The global warming trend observed over the last century, including an acceleration in warming since the mid-1900s, is well captured by most models. Many of the key spatial temperature patterns (Figure 5.1) and some important weather features like extratropical cyclones (§ 4.2.8) are also represented. However, systematic problems in temperature persist in some regions, and rainfall remains a significant modelling challenge and a major research priority, as models still fail to capture some key features of large-scale rainfall patterns.[44]

As we have seen throughout this book, the temporal variability of climate is crucial for anticipating and managing climate-sensitive health risks. The climate models can reproduce important features of sub-seasonal to year-to-year variability, particularly for temperature on large spatial scales. However, other features are poorly reproduced: for example, decadal variability in the Atlantic Ocean, which is important for climate prediction over the next 20–30 years.[44]

At the regional to sub-national scales of most interest to policy-makers, model performance is problematic. For example, the CMIP5 models reverse the observed relative intensities of Eastern Africa's two rainy seasons.[45] RCMs (see Section 9.3.1) are often used to add detail to coarse resolution GCM projections with the assumption that at least some of these global model biases are a result of regional-scale errors. However, these models inherit large-scale biases of the global model providing the boundary conditions, as well as adding their own assumptions and approximations with each step of processing. The Coordinated Regional Downscaling Experiment (CORDEX) has produced a wealth of downscaled climate-change projections that have been valuable for research purposes. However, it remains unclear how such experiments can be used in health applications because assessing confidence in regional model projections is even more challenging than for global models. High-resolution maps produced from these RCMs can give a misleading impression of high confidence in local climate-change impacts. These models and such products should not be used directly in decision-making without a thorough assessment of their strengths and weaknesses.

9.4.2 How accurate are the predictions?

In earlier chapters we discussed the importance of forecast skill (§§ 7.5 and 8.4) in understanding the utility of weather and climate information in decision-making. Skill can only be estimated given adequate samples of observations and historical or retrospective forecasts. For decadal and longer lead-times these data requirements are extensive, as the longer the lead-time of a prediction, the longer it takes to build up a sufficient sample to estimate the skill and reliability of the forecasts.

Currently, multi-annual (sometimes called 'decadal') prediction skill has only been assessed for lead-times of up to nine years.[46] Temperature forecasts at these lead-times have some skill when the forecasts are averaged over large regions, but not at the local levels desired for planning. Where the long-term temperature trends can be predicted with some skill, there is difficulty in predicting the natural variability around the trend.[41,47,48] In general, the temperature forecasts are best in regions where the externally forced trend is strong and there is minimal decadal variability about that trend.[46]

Skill for rainfall is marginal over most areas. Rainfall is more variable in both space and time than temperature (§ 5.2.5), and is less well simulated by climate models.[44] Although observed temperature trends are pronounced in most parts of the world, long-term trends in rainfall are, so far, undetectable above the background of year-to-year and decadal variability in most areas. In areas such as the Sahel, where there is significant decadal variability in rainfall, some of this variability may be predictable. At local scales, of most interest for societal impact, year-to-year fluctuations in rainfall can be substantial and rainfall projections are unlikely to be accurate.

Uncertainty in rainfall projections throughout the 21st century is dominated by this natural year-to-year and decadal variability and by climate model errors. For temperature, the importance of natural variability diminishes after a few decades, but model uncertainty causes substantial spread among longer-term projections. Divergence in anthropogenic emissions scenarios only becomes really important after several decades.[49,50]

The challenges of capturing decadal variability have significant implications for the development of long-term climate information services for the health sector. Planning cycles predominate at shorter timescales (see Chapter 3, Table 3.1), but even when longer-term information is desired it will likely be within the five to 20- or 30-year range when predicting the climate system is most difficult.

9.5 Conclusions

Changes in the patterns of weather and climate have already been observed and will become more pronounced as greenhouse gas concentrations in the atmosphere increase. Adaptation planning is therefore required to ensure that health systems are able to manage the impacts of the changing climate on public health.

Climate prediction on decadal and multi-decadal timescales has many challenges, particularly on the local and regional scales of most interest to policy and decision-makers. The cascading uncertainties of using outputs from such climate models to drive disease transmission models are significant impediments to developing robust predictions of specific diseases. However, the problems outlined in this chapter do not preclude any and all robust statements about the future impacts of climate change. Before using climate model outputs to assess future health impacts, it is essential to evaluate whether they are able to simulate the aspects of the climate system that are relevant to the particular health challenge being addressed. Where predictions are supported by physical understanding, and in cases where the observational record already shows evidence of local changes that corroborate predictions, we can place more confidence in those outcomes. As the impacts of climate change on the numerous pathways that influence disease transmission become detectable, a multidisciplinary approach to identifying vulnerabilities to climate change can be advanced.

Notes

i IPCC Fact Sheet: What is the IPCC? Available online at: www.ipcc.ch/news_and_events/docs/factsheets/FS_what_ipcc.pdf.

ii www.who.int/globalchange/resources/countries/en/.

References

1 WHO. *Report by the Secretariat* (WHO, Geneva, 2008).
2 Costello, A. *et al.* Managing the health effects of climate change. *Lancet* **373**, 1693–1733 (2009).
3 Markandya, A. *et al.* Health co-benefits from air pollution and mitigation costs of the Paris Agreement: a modelling study. *The Lancet Planetary Health* **2**, 126–133 (2018).
4 IPCC. Climate Change 2013: The Physical Science Basis. Contribution of Working Group I to the Fifth Assessment Report of the Intergovernmental Panel on Climate Change (eds. T.F. Stocker *et al.*).1535pp (Cambridge University Press, Cambridge, UK and New York, 2013).
5 Tyndall, J. On the absorption and radiation of heat by gases and vapours, and on the physical connexion of radiation, absorption, conduction. The Bakerian Lecture. Series 4, Vol. 22, pp. 169–194, 273–285. *The London, Edinburgh, and Dublin Philosophical Magazine and Journal of Science, Series 4* **22**, 169–194, 273–285 (1861).
6 Hulme, M. On the origin of'the greenhouse effect': John Tyndall's 1859 interrogation of nature. *Weather* **64**, 121–123, doi: 10.1002/wea.386View/save citation (2009).
7 Arrhenius, S. A. On the influence of carbonic acid in the air upon the temperature of the ground. *Philosophical Magazine and Journal of Science* **41**, 237–276 (1896).
8 Hawkins, E. & Jones, P. D. On increasing global temperatures: 75 years after Callendar. *Quarterly Journal of the Royal Meteorological Society* **139**, 1961–1963, doi:10.1002/qj.2178 (2013).
9 Keeling, C. D. *et al.* Atmospheric carbon dioxide variations at Mauna Loa Observatory, Hawaii. *Tellus* **26**, 538–551 (1976).
10 Easterling, D. R. *et al.* Observed variability and trends in extreme climate events: a brief review. *Bulletin of the American Meteorological Society* **81**, 417–425 (2000).
11 Sun, Y. *et al.* Impact of ocean warming on tropical cyclone size and its destructiveness. *Nature Scientific Reports* **7**, 8154 (2017).
12 Knutson, T. R. *et al.* Tropical cyclones and climate change. *Nature Geoscience* **3**, 157–163, doi:10.1038/NGEO779 (2010).
13 Vecchi, G. A. & Wittenberg, A. T. El Niño and our future climate: where do we stand? *WIRES Climate Change* **1**, 260–270, doi.org/10.1002/wcc.33 (2010).
14 Omumbo, J., Lyon, B., Waweru, S. M., Connor, S. & Thomson, M. C. Raised temperatures over the Kericho tea estates: revisiting the climate in the East African highlands malaria debate. *Malaria Journal* **10**, 12, doi:10.1186/1475-2875-10-12 (2011).
15 Shanks, G. D., Hay, S. I., Omumbo, J. A. & Snow, R. W. Malaria in Kenya's western highlands. *Emerging Infectious Diseases* **11**, 1425–1432. (2005).
16 Pascual, M., Ahumada, J. A., Chaves, L. F., Rodo, X. & Bouma, M. Malaria resurgence in the East African highlands: temperature trends revisited. *Proceedings of the National Academy of Sciences* **103**, 5829–5834 (2006).
17 Alonso, D., Bouma, M. J. & Pascual, M. Epidemic malaria and warmer temperatures in recent decades in an East African highland. *Proceedings of the Royal Society B: Biological Sciences* **278**, 1661–1669 (2011).
18 Ruiz D. *et al.* Multi-model ensemble (MME-2012) simulation experiments: exploring the role of long-term changes in climatic conditions in the increasing incidence of *Plasmodium falciparum* malaria in the highlands of Western Kenya. *Malaria Journal* **13**, 206, doi:10.1186/1475-2875-13-206 (2014).
19 Cash, B. A., Rodó, X. & Kinter, J. L. Links between Tropical Pacific SST and cholera Incidence in Bangladesh: role of the Western Tropical and Central Extratropical Pacific. *Journal of Climate* **22**, 1641–1660 (2009).
20 Rodo, X., Pascual, M., Fuchs, G. & Faruque, A. S. G. ENSO and cholera: a nonstationary link related to climate change? *Proceedings of the National Academy of Sciences* **99**, 12901–12906, doi:10.1073/pnas.182203999 (2002).
21 Hashizume, M. *et al.* The effect of rainfall on the incidence of cholera in Bangladesh. *Epidemiology* **19**, 103–110, doi:10.1097/EDE.0b013e31815c09ea (2008).

22 Pilson, R. D. & Leggate, B. M. A diurnal and seasonal study of the feeding activity of Glossina pallidipes. *Australian. Bulletin of Entomological Research* **53**, 541–550 (1962).
23 Hargrove, J. W. & Ackley, S. F. Mortality estimates from ovarian age distributions of the tsetse fly Glossina pallidipes Austen sampled in Zimbabwe suggest the need for new approaches. *Bulletin of Entomological Research* **105**, 294–304 (2015).
24 Van Ardenne, L., Wolski, P. & Jack, C. *Rekomitjie Climate Variability and Change. Cape Town: Climate Systems Analysis Group*. 96pp (University of Cape Town, Cape Town, 2017).
25 Vale, G. A., Hargrove, J. W., Cockbill, G. F. & Phelps, R. J. Field trials of baits to control populations of Glossina morsitans morsitans Westwood and G. pallidipes Austen (Diptera: Glossinidae). *Bulletin of Entomological Research* **76**, 179–193 (1986).
26 Hargrove, J. W. in *The Trypanosomiases* (eds. I. Maudlin, P.H. Holmes, & M.A. Miles), 113–137 (CABI Publishing, Wallingford, 2004).
27 Greene, A. M., Goddard, L. & Cousin, R. Web tool deconstructs variability in twentieth-century climate. *EOS Transactions American Geophysical Union* **92**, 397, doi:10.1029/2011EO450001 (2011).
28 Greene, A. M., Goddard, L., Gonzalez, P. L., Ines, A.V. & Chryssanthacopoulos, J. A climate generator for agricultural planning in southeastern South America. *Agricultural and Forest Meteorology* **203**, 217–228, doi:10.1016/j.agrformet.2015.01.008 (2015).
29 Muñoz, A. G., Thomson, M. C., Goddard, L. & Aldighieri, S. Analyzing climate variations at multiple timescales can guide Zika virus response measures. *GigaScience* **20165**, 41, doi:10.1186/s13742-016-0146-1 (2016).
30 Williams, A. P. & Funk, C. A westward extension of the warm pool leads to a westward extension of the Walker circulation, drying eastern Africa. *Climate Dynamics* **37**, 2417–2435, doi:10.1007/s00382-010-0984-y (2011).
31 Collins, M. E. A. Long-term Climate Change: Projections, Commitments and Irreversibility. Climate Change 2013: The Physical Science Basis. Contribution of Working Group I to the Fifth Assessment Report of the Intergovernmental Panel on Climate Change, 1029–1136 (2013).
32 Lyon, B. & DeWitt, D. G. A recent and abrupt decline in the East African long rains. *Geophysical Research Letters* **39**, L02702, doi:10.1029/2011GL050337 (2012).
33 Mohan JE *et al.* Biomass and toxicity responses of poison ivy (Toxicodendron radicans) to elevated atmospheric CO2. *Proceedings of the Nattional Academy of Sciences* **103**, 9086–9089 (2006).
34 Myers, S. S. *et al.* Increasing CO_2 threatens human nutrition. *Nature* **510**, 139–142, doi:10.1038/nature13179 (2014).
35 Neumann, B., Vafeidis, A. T., Zimmermann, J. & Nicholls, R. J. Future coastal population growth and exposure to sea-level rise and coastal flooding – a global assessment. *PLoS One* **10**, e0118571, doi:10.1371/journal.pone.0118571 (2015).
36 Leatherman, S. P. & Beller-Simms, N. Sea-level rise and small island states: an overview. *Journal of Coastal Research Special Issue: Island States at Risk. Global Climate Change, Development and Population* **24**, 1–16 (1997).
37 Tate, R. D., Benkendorff, K., Lah, R. A. & Kelahera, B. P. Ocean acidification and warming impacts the nutritional properties of the predatory whelk, *Dicathais orbita. Journal of Experimental Marine Biology and Ecology* **493**, 7–13 (2017).
38 Sellers, S. & Ebi, K. Climate change and health under the Shared Socioeconomic Pathway Framework. *International Journal of Environment and Public Health* **15**, pii: E3, doi:0.3390/ijerph15010003. (2017).
39 Trzaska, S. & Schnarr, E. *A Review of Downscaling Methods for Climate Change Projections.* 1–42 (USAID, Washington, DC, 2014).
40 Trzaska, S. & Schnarr, E. *A Review of Downscaling Methods for Climate Change Projections.* 42pp (USAID, Washington, DC, 2014).
41 Goddard, L. *et al.* A verification framework for interannual-to-decadal predictions experiments. *Climate Dynamics* **40**, 245–272 (2013).
42 Carlson, D., Eyring, V., van der Wel, N. & Langendijk, G. *WCRP's Coupled Model Intercomparison Project: A Remarkable Contribution to Climate Science* (EGU, Munich, 2017).

43 Nissan, H., Ukawuba, I. & Thomsom, M. C. *Factoring Climate Change into Malaria Eradication Strategy*. 63pp (GMP WHO, Geneva, 2017).
44 Flato, G. E. A. in *Climate Change: The Physical Science Basis. Contribution of Working Group I to the Fifth Assessment Report of the Intergovernmental Panel on Climate Change* (ed. T.F. Stocker *et al.*), 741–882 (Cambridge University Press, Cambridge, 2013).
45 Lyon, B. & Vigaud, N. in *Climate Extremes: Patterns and Mechanisms* (eds. S-Y.S. Wang, J-H. Yoon, C.C. Funk, & R.R. Gillies), 265–281 (John Wiley & Sons, Hoboken, NJ, 2017).
46 Meehl, G. A. E. A. Decadal climate prediction: an update from the trenches. *Bulletin of the American Meteorological Society* **95**, 243–267 (2014).
47 Doblas-Reyes, F. J. *et al.* Initialized near-term regional climate change prediction. *Nature Communications* **4**, 1715, doi:10.1038/ncomms2704 (2013).
48 van Oldenborgh, G. J., Doblas-Reyes, F. J., Wouters, B. & Hazeleger, W. Decadal prediction skill in a multi-model ensemble. *Climate Dynamics* **38**, 1263–1280 (2012).
49 Hawkins, E. & Sutton, R. The potential to narrow uncertainty in regional climate predictions. *Bulletin of the American Meteorological Society* **90**, 1095–1107 (2009).
50 Hawkins, E. & Sutton, R. The potential to narrow uncertainty in projections of regional precipitation change. *Climate Dynamics* **37**, 407–418 (2011).

10

CLIMATE INFORMATION FOR PUBLIC HEALTH ACTION

Challenges and opportunities

Madeleine C. Thomson and Simon J. Mason
Contributors: John del Corral, Andrew Kruczkiewicz, Gilma Mantilla and Cristina Li

> Water, water everywhere,
> Nor any drop to drink
>
> 'The Rime of the Ancient Mariner' *by Samuel Taylor Coleridge*

10.1 Introduction

Throughout history, weather has played a major role in all aspects of human endeavour including the outcome of wars and battles. A Japanese word that has been adopted into English is even derived from such instances – 'kamikaze'. In 1274 and 1281 Kublai Khan led two invasions of Japan, but both were thwarted by what seemed like miraculously timed typhoons (§ 4.2.8). The typhoons were then denoted 'divine winds' (or 'kamikaze'). War has also played a major role in the development of weather forecasting. For example, the modern numerical approach to weather forecasting (§ 7.4) was initiated by an experiment conducted in World War I. Further major improvements in technology and observation were made during World War II. It should therefore come as no surprise that many meteorological agencies were initially established by the military. The UK Met Office, for example, was within the Ministry of Defence until 2011 when it was reorganized as part of the Department for Business, Innovation and Skills.

While weather and seasonal climate conditions may impact military capability and operational effectiveness, climate change is identified as a national security issue[1] and as a threat to global security.[2] Health security (see § 1.3.8) requires a cross-sectoral approach, and brings together a wide range of government agencies including national security, health, agriculture, environment wildlife, communication, etc. to tackle imminent health threats.[3] The capacity of the National Meteorological and Hydrological Services (NMHS) to contribute to health security is being strengthened through new investments in hydro-meteorological services.[4] The threat that both climate variability and change pose to the achievement of development targets has resulted in an increasing concern for climate-sensitive development sectors, including health. Collaborative platforms have emerged involving government, academia, civil society and the private sector to better understand, mitigate and manage climate-related risks to health.[5] While these new efforts are welcome it is clear that health lags behind other sectors in responding to climate risks and the potential use of climate information to help reduce those risks.[6]

Throughout this book, we have sought to identify how, when and where climate information, based on historical data, monitoring products and predictions of future weather and climate can be used to inform health policy and practice. We have prioritized operational information over research opportunities to ensure a focus on practical outcomes. In this chapter, we provide a short review of climate information currently available, as described in detail in Chapters 4–9. We then explore how developments in technologies, institutional arrangements and the education of health professionals are providing new opportunities for translating climate information into a new resource for health-sector decision-making.

10.2 Climate services for health

There are numerous opportunities for creating climate information that can serve to improve health decision-making (Table 10.1).

These include creating information and services from detailed historical analyses as the basis for risk assessment. For example, a review of prior years can help identify at what time of year heat waves are most likely to occur and health services should be on high alert, or how consistent the onset of the malaria transmission season is and when control programmes should initiate indoor residual spray campaigns. With sufficient historical data (80 years) it is possible to distinguish year-to-year and decadal variations as well as long-term trends in the climate. These analyses can help identify for example, whether or not a health event is triggered by El Niño or other global climate predictors.

Monitoring products, using a selection of satellite, model and ground observations (Chapter 6), provide users with near real-time climate information that may be predictive of some health events months in advance. Historical and monitoring information are available for all regions of the world and in all seasons whether or not weather and climate forecasts are available. However, the value of the information products depends on the quality of the data relative to the spatial and temporal scale of the problem being addressed and whether or not they are available to the user in near real-time. All climate information is compromised if its construction relies on sparse and poor quality historical data, as is common in many developing countries. There are significant opportunities for improving national climate information in developing countries if the best globally available products (e.g., satellite or reanalysis) are combined with quality-assessed local ground observations from the national climate archives and monitoring services.[7] Initiatives to rescue

TABLE 10.1 Summary of climate information of relevance to health sector decision-making

Weather- and climate-informed intervention	*Climate information*	*Data requirements*	*Where practical*	*When practical*	*Chapters*
Risk assessment for better targeting of interventions					
Assessment of underlying drivers of climate risk to health	Timescale decomposition of variability in historical climate data into interannual, decadal and long-term change	80 years monthly data	Wherever data quality permits	All year, but best to analyse by season	2,3,5
Routine spatial targeting of at-risk population	Indication of geographic region where climate suitability favours particular health risk	Historical climatological data for specific risk indicator – with sufficient years of data, estimates of uncertainty can be included		Whenever data quality permits	2,3,6
Routine seasonal targeting of interventions	Indication of time of year where climate suitability favours health risk	30 years daily data permits detailed characterization of seasonality of extremes, onset, offset and peak season, monthly data for shorter time series can be used for simpler analyses	Wherever data quality permits	All year	2,3,6
Early warning of emergent health risks					
Assessment of interannual variations in health risk associated with ENSO (or other predictors) and therefore potentially predictable using a seasonal climate forecast	ENSO (or other predictor) impact assessment	30 years of monthly or seasonal climate data along with SST data (e.g., ONI)	Wherever data quality permits and there is a defined ENSO (or other predictor) effect	During seasons where there is a defined ENSO (or other) effect	2,3,5,8

(*Continued*)

Early warning of adverse health events based on current and cumulative climate observations	Near real-time monitoring of climate variables and model that links climate variables to health outcome	Daily, decadal or monthly climate products	Wherever there are reliable near real-time monitoring products (plenty for rainfall but not for temperature)	All year	2,3,6
Early warning of onset or offset in climate suitability for seasonal disease transmission	Weather forecasting up to ten days	Daily averaged weather forecasts	Most skilful in the extra-tropics	Before the onset, offset	2,3,7
Early warning of hot or cold spells	Weather forecasting up to ten days	Daily averaged weather forecasts	Most skilful in the extra-tropics	Most skilful in the winter	2,3,7
Early warning of future winter storms and tropical cyclones with immediate health threats	Weather extremes – e.g., tropical storm forecasts up to ten days	Hourly to daily averaged weather forecasts	In the extra-tropics or in tropical storm-affected regions	During winter storm or tropical cyclone seasons	2,3,4,7
Early warning of health risks one week to one month in advance	Sub-seasonal forecasting	Averaged weather forecasts (not yet operational)	Still experimental	Most skilful in the winter season but limited testing	2,3,7
Early warning of seasonal changes in health risks	Forecasting of seasonally-averaged climate – integrated with climate monitoring system	Seasonal forecasts as seasonally-averaged probabilities	In the tropics	Most skilful during seasons where there is a defined ENSO (or other) effect	2,3,8
Early warning of interannual changes in health risks	Forecasting interannual climate within the next few years	Interannual forecasts as annually or multi-annually averaged probabilities	Still experimental	Most skilful in years two and three and with temperature	2, 3, 9

(Continued)

TABLE 10.1 (Continued)

Weather- and climate-informed intervention	*Climate information*	*Data requirements*	*Where practical*	*When practical*	*Chapters*
Long term planning and preparedness – for adaptation					
Interventions based on decadal and multi-decadal forecasting	Decadal and multi-decadal forecasting	Decadal prediction – next two to nine years	Low skill everywhere	Not yet viable	2,3,9
Early warning of shifts in climate averaged over 30 years for adaptation planning	Multi-decadal/climate change	10.2.2 Multi-decadal temperature and rainfall averaged over 30 years; best where decadal variations are weak or absent	Globally for temperature; rainfall limited to specific regions	All year, but best analysed by season	2,3,9
Use of climate in assessment of the impact of interventions					
Assessment of impact of climate-sensitive health interventions (removing the climate component to reveal the intervention component)	Analysis of historical climate data to assess difference between base line and intervention years	Data consistent with spatial and temporal characteristics of health data	Wherever data quality permits	All year	2,3,6

Abbreviations: ENSO, El Niño – Southern Oscillation; ONI, Oceanic Niño Index; SST, Sea Surface Temperature.

observational data and to work with NMHS to integrate all relevant station data with global products (see Box 6.2) are essential to improve the quality and coverage of climate data needed for decisions. Sometimes the lack of availability of national observations is the result of a data gap; sometimes it may be the result of NMHS policy that provides forecast information products freely to end-users but withholds access to high-resolution historical data. Data policies vary by country, but whatever the underlying reason for the policy, lack of ready access to quality historical data limits user engagement in climate services.

The emergence of climate services over the last decade has included a strong focus on forecasts and projections of weather and climate.[8,9] In Chapters 7, 8 and 9 we describe the basis of these predictions and when and where they are likely skilful. All forecasts vary in their skill depending on the lead-time, timescale, spatial scale, geographic region, season and over years and decades. What works in in one region or season may not work well in another. The use of climate and weather forecasts has to take into account this complexity, and so their relevance for specific health issues must be tested locally with the help of relevant experts. Standard operating procedures (SOPS) for the development of an international response to El Niño – Southern Oscillation (ENSO) forecasts have recently been developed.[10] These SOPS will need to be elaborated carefully at the national and subnational level to enable an effective response.

The weather and climate forecasting community are pushing the boundaries of predictability at all timescales. Having demonstrated predictability and utility at weather, seasonal and climate-change timescales they are increasingly paying attention to the in-between timescales – sub-seasonal and decadal – in the search for predictability (see Box 7.4 and § 9.3.2). Indications are that sub-seasonal predictions may have a role to play in health early warning systems in the coming years.[11]

Enabling the development and uptake of climate services are: i) major advances in technology that can connect data on the one hand and people on the other; ii) changes in institutional arrangements that support multi-sectoral collaborations; and iii) a growth of educational and professional training initiatives that support the health communities' understanding and use of climate knowledge and information. These areas are elaborated further below.

10.3 Advances in technology

The evolution of our capacity to monitor and predict the climate has been enabled through advances in information and communication technologies (ICT). Advances in recent decades have had a profound impact on societies, transforming many to knowledge-based economies where the ability to integrate disparate information is central to making more effective decisions. These technological advances have increased the connectedness between data and people. This increase in connectedness is substantially due to massive increases in computer power and data storage as well as the global penetration of ICT technologies. A summary of

BOX 10.1 TECHNOLOGY CHANGES PRE, DURING AND POST THE MDG ERA

John del Corral and Andrew Kruckiewicz, IRI, Columbia University, New York, USA

The pre-MDG era (1970–2000)

In the pre-MDG period, professionals from both climate and health communities were working on desktop PCs and mainframes. Information was shared via FTP, email, bulletin boards, floppy disks and large magnetic tapes. The emergence of the internet as a research tool in the 1980s and the arrival of the World Wide Web (WWW) as a publicly available communication service in 1989 created a phenomenal new capacity to connect data and information on the one hand and people on the other (see Figure 10.1).

This early stage of development of the WWW is often referred to as Web 1.0.[12] Research and governmental institutions began creating their own websites for the online dissemination of knowledge and data.

The rapidly developing internet was quickly exploited by initiatives, such as the Program for Monitoring Emerging Diseases (ProMED), designed to enable prompt reporting of disease outbreaks.[13] Thus, by the beginning of the new millennium, an extraordinary set of new capacities was available to the health community to understand, predict and manage environmentally-determined health risks. However, many of these opportunities were limited to technically savvy individuals in government research laboratories in the developed world. In1995 it was estimated that only 0.4% of the global population had access to the internet.

MDG era (2000–2015)

Prior to the turn of the millennium the majority of research articles were published in hard copy by scholarly societies for their members (for a fee) and accessed by students and researchers via university libraries; a service severely limited in many developing countries. The arrival of the WWW revolutionized the way research literature could be found, accessed and consumed. With increasing penetration of the internet, students and scholars around the world could download the material they needed from online sources and Open Access (OA) journals, such as the *Malaria Journal* (where papers are available online and free of charge to the user).[14] The Social Web (referred to as Web 2.0) arrived in the late 1990s and early 2000s enabling social networks, blogging and wikis. Now, a new generation of information gatherers, providers and collaborators could contribute to the global knowledge-base.

More astoundingly, mobile broadband technology grew from 738 million mobile cellular subscriptions in 2000 to more than seven billion by the end of the MDG era.[15] The technology provided developing countries with the opportunity to leapfrog cumbersome fixed telephone and broadband technologies, and mobile phones emerged as the communication method of choice for disaster alerts.[16]

With the advent of *big data* analytics it became possible to mine millions of data streams (including electronic health patient records, social media, internet, mobile phones and remote sensing) in an instance.[17] This capability was exploited by Google who monitored online behaviour to provide predictions of flu trends based on an algorithm that captured search behaviour that was initially highly correlated with patients presenting with flu symptoms. The preliminary results from the Google Flu Trends (GFT) tool looked promising. Not only did the GFT predict flu outbreaks ten days before they were observed by the Center for Disease Control, but they were considered to be potentially more accurate.[18] However, after GFT predicted flu trends poorly for a number of years, it became clear that this analytical approach cannot supplant more traditional methods of data collection and analysis.[17] GFT algorithms are not open to public scrutiny, which raises broader concerns of the role of private companies in providing public predictions of epidemics.

ICT may also play a critical role in educating the global public health community about the risks of climate variability and change to health outcomes and the opportunities for improved health created through climate information. The development and provision of Massively Open Online Courses (MOOCs) has expanded rapidly in the past ten years, and their use in climate and health education and profession training is increasing (see Table 10.2).

Sustainable Development Goals era (2015–2030)

The latest evolution of the Web is Web 3.0, or the Semantic Web. This phase emerged towards the end of the MDG period through the realization of 'linked data' on the web. Online knowledge and data can now be classified in ontologies (a set of concepts and categories in a subject area or domain that shows their properties and the relations between them) with inference rules. Sophisticated semantic queries can be applied to this information to extract related content from large databases. These queries can lead to the discovery of relationships between pieces of information that had not been thought of or identified before. Intelligent 'guided' searches can be applied to online data that have semantic metadata. This new capacity is a significant opportunity for data integration in the time of the Sustainable Development Goals (SDGs).

TABLE 10.2 Resources in the climate change–climate and health arena

- Climate Literacy – http://flexible.learning.ubc.ca/news-events/climate-literacy-mooc-launches/
- Climate Services – www.kiusdesign.com.ar/worldBank/index.html
- Climate Change and Health – https://iversity.org/en/courses/climate-change-and-health
- Climate Change Policy and Public Health – https://moocs.wisc.edu/mooc/climate-change-policy-and-public-health/
- Climate Change and Health US-Gov – https://health2016.globalchange.gov/
- Climate Change and Health – www.unitar.org/new-climate-change-learning-modules-health-and-cities-now-available-online
- Climate Curriculum, Climate Information for Public Health Short Course – https://academiccommons.columbia.edu/catalog/ac:130711
- Global Consortium on Climate and Health Education – https://www.mailman.columbia.edu/research/global-consortium-climate-and-health-education

The SDGs call for an integrated approach to development at the country level, and a set of ambitious targets that will be achieved by 2030 (see Chapter 1). The goals include an end to extreme poverty and hunger, while improving access to health care and education, protecting the environment, and building peaceful, inclusive societies. Many of the technological trajectories that emerged during the MDG period are likely to continue during the SDGs. Increasing access to information technologies and massive increases in access to data at higher and higher spatial and temporal resolutions and increasingly in near real-time will certainly continue. According to the 2017 International Telecommunications (ITU) report,[i] 80% of global youth now have access to the internet in 104 countries. Not surprisingly, data and ICT have been prioritized as a core resource to deliver the SDGs (including SDG 13 'to take urgent action on climate change and its impacts'), and to measure progress to their achievement.[19]

the evolution of ICT, from before the Millennium Development Goals (MDG) era to the present day, illustrates the speed of change and is provided in Box 10.1.

While technologies have, and continue to, transform our lives at a phenomenal rate, health decision-makers and the public have struggled to keep up with the sheer volume of data, approaches and tools becoming available. In part, this is because of a proliferation of information services that are not specifically tailored to health user needs. Web platforms designed to support specific health decision-makers are often created by researchers without the capacity (resourcing or interest) to deliver routine dynamic information for specific decision-making needs in a sustainable manner. For example, of 11 web-based platforms designed to service malaria decision-makers only two (the IRI Malaria Maproom and Health

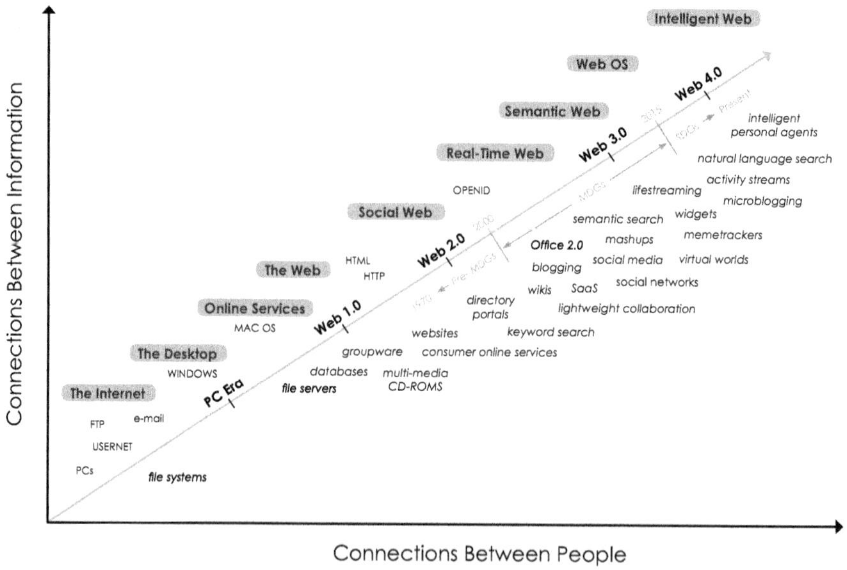

FIGURE 10.1 Connecting information and people. Image adapted with permission from "the intelligence is in the connections" created by Nova Spivack www.novaspivack.com

Map) were automatically updated in near real-time.[20] For national-level planning and surveillance, the District Health Information System II (DHIS2) software has emerged as a global leader in the development of digital tools for the health sector in developing countries[21] and efforts are underway to integrate climate information into the DHIS2.

Technological advances in supercomputing and data storage are also helping to refine details in climate models and providing increasingly accurate early warnings of weather and climate risks at multiple timescales (Chapters 6, 7, 8 and 9). At all forecasting timescales beyond a few hours or days, uncertainty is substantial. Reducing model uncertainties is a priority for the weather and climate community. Investments in a new generation of climate models, capable of resolving finer details, including cloud systems and ocean eddies, is justified given the opportunity to improve forecasts of extreme events and reduce uncertainty in climate-change projections used for adaptation purposes.[22] Technology also plays a major role in enhancing the dissemination and uptake of climate information by decision-makers and the public. However, while technologies bring enormous opportunities they do not serve all equally and they may also bring new and unprecedented risks to society.[23]

10.4 Institutional arrangements

The landscape of actors involved in climate and health activities is broad. It includes United Nations (UN) and multi-lateral agencies, many key institutions representing civil society, health policy and practitioner organizations, government and partners from other sectors, private industry, non-governmental organizations, academics, philanthropists and development banks. Some of the key drivers of international health policy are described in Chapter 1. Increasing alignment of policy processes around the SDGs provides new opportunities for integrating climate into health policy-making.

The basic building blocks of climate and health interactions revolve around four levels of public health authority: international, national, sub-national and local. International collaboration in infectious disease is particularly important since diseases do not respect national boundaries, and pandemics pose health security threats to far beyond their country of origin. The national level is where policies, developed at home or internationally, are implemented through further elaboration and legislation. Regional authorities often play important roles in translating national policies to local realities. Finally, public health actions, both in developed and undeveloped countries, substantially happen at the local level where regional and local health authorities, private concerns and civil society organizations have sufficient direct access to the population. Decentralization of decision-making to regional bodies provides new opportunities for local meteorological services to engage with local stakeholders. For example, in Kenya significant devolution of government authority to the county (regional level) has occurred in the last five years. Directors of 47 county meteorological services are now empowered to work directly with county-level health directorates providing new opportunities for local innovation.

Health care financing may influence the type and effectiveness of health-related climate-change adaptation and mitigation policies at the national level. During the MDG era, development assistance to lower-income developing countries has prioritized commodities (drugs and medical supplies) for targeted diseases rather than the delivery of health care services and the monitoring of the effectiveness of disease programmes. Without a strong health system with an effective supply-chain management, it is hard to see how climate information can be used to improve the delivery and targeting of health services. The countries most at risk of climate-related health challenges are in Africa and South East Asia where health systems are especially weak. In 2014, only a small proportion (2.84%) of official health development assistance to the 20 countries most vulnerable to climate change focused on the health effects of climate change.[24] Development assistance for strengthening health systems could be reinforced if climate change adaptation funding (e.g., from the Green Climate Fund) were also used to strengthen health care delivery through reductions in climate-related health risks.

In the last few years, the climate and health communities have been coming together in an unprecedented way to ensure the development of weather and climate services that can truly meet the needs of the health community. However, lack of evidence of climate risks to health and the value of climate information to health decision-making in specific contexts is a significant barrier to greater investment in climate and health research. To date, the extensive resources in weather and climate data have been little used in practical decision-making. When investments are made, converting scientific knowledge to information suitable for uptake by policy-makers is critical if results are to be translated into practical outcomes.[25] Decisions occur at the interface of a number of perspectives including those provided by policy and stakeholder communities. Decisions are also constrained by organizational perspectives (such as the timing when decisions are routinely made or when budgets are released), and build on practitioner experience and capacities (Figure 10.2). Thus, mapping institutional mandates, national and subnational policies and practices along with local capacities (and local champions) and resource flows is a necessary starting point for the development of climate services for the community.

10.5 Education and training

One way to increase uptake of research evidence in policy and practice is to ensure that climate-change issues are taught in schools of medicine, public health, nursing, etc. This will ensure that health workers are familiar with the knowledge, data, methodologies and tools necessary to incorporate climate into health decisions.

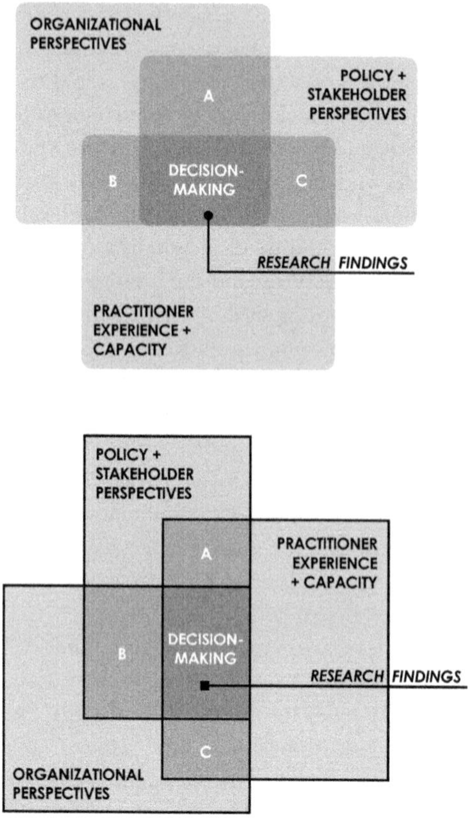

FIGURE 10.2 Research findings can better inform decision-making if they take into account the different perspectives that influence decision-making

Over 2400 years ago the Greek physician, Hippocrates, *Father of Modern Medicine*, encouraged his medical students to understand the local environmental characteristics and to consider seasonal and unusual climatic factors when diagnosing disease in their patients.[26] This perspective has been maintained over the millennia through various schools of medicine, and remains relevant today.[27] The emergence of health professionals as a trusted voice in alerting the public to climate-change risks as well as the health co-benefits of climate-change mitigation has reinforced the need for a climate-literate health community,[28] and the introduction of climate-change education into global public health.[29,30] Some medical institutions are starting to include limited modules on climate change in the core curricula of undergraduate and graduate programmes, but much more needs to be done (Case Study 10.1).

CASE STUDY 10.1 CLIMATE AND HEALTH EDUCATION IN COLOMBIA

Gilma Mantilla, Universidad Javeriana, Colombia, and Cristina Li, Mailman School of Public Health, Columbia University, New York, USA

The Pontificia Universidad Javeriana undertook a survey in October 2017 to evaluate the current state of climate-change and health education in universities across Colombia. The ultimate objective was to better understand how education can be improved to help build the next generation of doctors that are cognizant of climate-change issues and to create a healthier and more secure future.

The survey comprised of 28 questions presented in three parts: part one, titled 'Identification', with demographic questions; part two, titled 'Climate Change and Health Courses', with questions including types of courses offered, number of students enrolled, professor profile, types of resources utilized and barriers to implementing courses; and part three, titled 'Projects and Research on Climate Change and Health', with questions pertaining to financing and objectives of such research. The survey was then distributed by e-mail to all universities in Colombia with an accredited Department of Medicine.

Fifty-nine universities in Colombia were identified with Departments of Medicine. Out of these universities, a total of 47 (80%) responded, across 17 states. The survey revealed that 25 universities offer the topic of climate change and health as part of their undergraduate programme, with only Pontificia Universidad Javeriana in Bogotá offering the topic at both the undergraduate and graduate level. Twenty-one of these universities included a course on climate change within a mandatory course of public health or epidemiology in their curriculum.

Climate-change and health sessions have been taught in one or more universities for the last eight years. Across all institutions, an average of

60 students are currently taught per semester. The most common resource used to teach climate-change and health sessions is scientific literature, followed by documentaries and movies, internet and web pages, and newspapers and print articles.

Out of 22 (47%) universities that reported that they did not have climate-change and health education, 15 have not attempted to implement any course into their curriculum, with the biggest barriers including a lack of personnel to develop the curriculum and a lack of available hours to include the course within the curriculum. As of 2017, only four universities in Colombia have initiated research projects on climate change and health, suggesting that the priority of climate-change and health education is low among the medical community.

Although universities in Colombia have included the topic of climate change and health as a session in a regular course of public health for undergraduate students for an average of eight years, it is clear that there is still more that needs to be done in emphasizing the importance of the topic for future generations of students. To meet this need, medical schools in Colombia must continue to focus their broader curricula around the intersection of climate change and social determinants of health. They must continue to develop research and training to increase the health community's capacity to understand, use and demand appropriate climate information, which is of primary importance to understand how to manage climate-related diseases and how to aid not only the communities they serve, but also the planet.

New initiatives are underway, such as the Global Consortium on Climate and Health Education (GCCHE), which aims to unite medical schools, nursing schools and schools of public health in sharing best practices to build curricula and core training.[31]

The risks to humanity of exceeding planetary boundaries are increasingly clear.[32] Health professionals are now looking to connect large-scale biophysical and societal processes to what is happening to populations within their area of concern. Once near-term climate-related risks to health have been identified it is possible to seek local solutions. Thus, when considering how we might adapt to a changing climate, we are forced once more to reconnect with the Hippocratic perspective and focus our attention on understanding the local climate, its drivers, seasonality, variability (including extremes) and longer-term trends.[33] The reader is directed to 'Climate Services for Health – Improving Public Health Decision-making in a New Climate'. This guide seeks to promote the development of climate services for the health sector.[34] It advocates the creation of tailored climate services by building an enabling environment through: guaranteeing sufficient human and technical capacity; compiling and conducting necessary research; and undertaking purpose-driven product and service development. Climate services are poised to offer a critical contribution to society's capacity to protect health from emerging climate-related risks. As evidence of the increasing interest from the health community, the Lancet Countdown[33] will report annually on a set of indicators that reflect progress on health and climate change, including the development of climate services for health decision-making.

10.6 Conclusions

This book is based on the premise that health professionals, researchers and policymakers need increasingly to take climate into account when promoting policy, implementing programmes and treating patients. To work with climate information effectively they need to become climate-literate, i.e., to have enough basic knowledge on the climate system, how it is measured and modelled, to be able to make the correct inference from climate information (historical, current and future), and to respond accordingly. They should not travel this road alone; this book seeks to be a valued companion and is designed to stimulate further dialogue and collaboration between health specialists and climate information providers.

Note

i See ITU report. Available at: https://www.itu.int/en/ITU-D/Statistics/Documents/facts/ICTFactsFigures2017.pdf.

References

1 Morales Jr., E. Global climate change as a threat to U.S. national security. *Journal of Strategic Security* **8**, 133–148 (2015).
2 WEC. *The Global Risks Report 2016* (World Economic Forum, Geneva, 2016).
3 Kandel, N. *et al.* Joint external evaluation process: bringing multiple sectors together for global health security. *The Lancet Global Health* **5**, e857–e858, doi:https://doi.org/10.1016/S2214-109X(17)30264-4 (2017).
4 Rogers, D. P. & Tsirkunov, V.V. *Weather and Climate Resilience Effective Preparedness through National Meteorological and Hydrological Services* (Development/The World Bank, Washington, DC, 2013).
5 Vaughan, C. & Dessai, S. Climate services for society: origins, institutional arrangements, and design elements for an evaluation framework. *Wiley Interdisciplinary Reviews. Climate Change* **5**, 587–603, doi:10.1002/wcc.290 (2014).
6 Jancloes, M. *et al.* Climate Services to improve public health. *International Journal of Environment and Public Health* **11**, 4555–4559 (2014).
7 Dinku, T. *et al.* *THE ENACTS APPROACH: Transforming Climate Services in Africa One Country at a Time* (World Policy Institute, New York, 2016).
8 Hewitt, C., Mason, S. & Walland, D. The global framework for climate services. *Nature Climate Change* **2**, 831–832 (2012).
9 Brasseur, G. P. & Gallardo, L. Climate services: lessons learned and future prospects. *Earth's Future* **4**, 79–89, doi:002/2015EF000338 (2016).
10 IASC. *Inter-Agency SOPs for Early Action to El Niño/La Niña Episodes.* 35pp (Inter-Agency Standing Committee for the UN, Geneva, 2018).

11 Tompkins, A. M. *et al.* Predicting climate impact on health at sub-seasonal and seasonal timescales, in *Sub-seasonal to Seasonal Prediction: The Gap Between Weather and Climate Forecasting* (eds. A. Robertson & F. Vitart) (Elsevier, forthcoming 2018).
12 Choudhury, N. World Wide Web and its journey from Web 1.0 to Web 4.0. *International Journal of Computer Science and Information Technologies* **5**, 8096–8100 (2014).
13 Yu, V. L. & Madoff, L. C. ProMED-mail: an early warning system for emerging diseases. *Clinical Infectious Diseases* **39**, 227–232, doi:10.1086/422003 (2004).
14 Smith, E., Haustein, S., Mongeon, P., Shu, F. & Ridde, V. Knowledge sharing in global health research – the impact, uptake and cost of open access to scholarly literature. *Health Research Policy and Systems* **15**, 73, doi:10.1186/s12961-017-0235-3 (2017).
15 UN. *The Millennium Development Goals Report*. 72 (United Nations, New York, 2015).
16 Chan Mow, I., Sasa, H., Shields, C. & Fitu, L. Towards a people centered early warning and disaster response system in Somoa: the use of ICT by Somoans during disaster. *The Electronic Journal of Information Systems in Developing Countries* **81**, 1–18 (2017).
17 Bansal, S., Chowell, G., Simonsen, L., Vespignani, A. & Viboud, C. Big data for infectious disease surveillance and modeling. *The Journal of Infectious Diseases* **214**, S375–S379, doi:https://doi.org/10.1093/infdis/jiw400 (2016).
18 Ginsberg, J. *et al.* Detecting influenza epidemics using search engine query data. *Nature* **457**, 1012–1014, doi:10.1038/nature07634 (2009).
19 Earth Institute and Ericsson. *How Information and Communications Technology can Achieve the Sustainable Development Goals: ICT and SDGs* (2016). Available at: http://unsdsn.org/wp-content/uploads/2015/09/ICTSDG_InterimReport_Web.pdf.
20 Briand, D., Roux, E., Desconnets, J. C., Gervet, C. & Barcellos, C. From global action against malaria to local issues: state of the art and perspectives of web platforms dealing with malaria information. *Malaria Journal 17*, 122 (2018).
21 Kiberu, V. M. *et al.* Strengthening district-based health reporting through the district health management information software system: the Ugandan experience. *Medical Informatics and Decision Making* **14**, 40, doi:10.1186/1472-6947-14-40 (2014).
22 Palmer, T. Climate forecasting: build high-resolution global climate models. *Nature* **515**, 338–339, doi:10.1038/515338a (2014).
23 WEC. *The Global Risks Report 2017* (World Economic Forum, Geneva, 2017).
24 Gupta, V., Mason-Sharma, A., Caty, S. N. & Kerry, V. Adapting global health aid in the face of climate change. *The Lancet Global Health* **5**, 133–134 (2017).
25 Ramirez, B. & TDR-IDRC Research Initiative on Vector Borne Diseases and Climate Change. Support for research towards understanding the population health vulnerabilities to vector-borne diseases: increasing resilience under climate change conditions in Africa. *Infectious Diseases of Poverty* **6**, 164, doi:10.1186/s40249-017-0378-z (2017).
26 Clifton, F. *USC School of Medicine Digital Collections* (London, 1734).
27 Valencius, C. B. Histories of medical geography. *Medical History Supplement*. **20**, 3–28 (2000).
28 Rudolph, L. & Harrison, C. A. *Physician's Guide to Climate Change, Health and Equity* (Public Health Institute, Oakland, CA, 2016).
29 Maxwell, J. & Blashki, G. Teaching about climate change in medical education: an opportunity. *Journal of Public Health Research* **5**, 673, doi:10.4081/jphr.2016.673 (2016).
30 Friedrich, M. J. Medical community gathers steam to tackle climate's health effects. *Journal of the American Medical Association* **317**, 1511–1513 (2017).
31 Shaman, J. & Knowlton, K. The need for climate and health education. *American Journal of Public Health* **108**, S66–S67, doi:10.2105/AJPH.2017.304045 (2017).
32 Steffen, W. *et al.* Planetary boundaries: guiding human development on a changing planet. *Science* **347**, 1259855 (2015).
33 Watts, N. *et al.* The Lancet Countdown: tracking progress on health and climate change. *The Lancet* **389**, 1151–1164 (2017).
34 Shumake-Guillemot, J. & Fernandez-Montoya, L. *Climate Services for Health: Improving Public Health Decision-making in a New Climate* (WMO/WHO, Geneva, 2016).

INDEX

Action and Investment to Defeat Malaria, 2016–2030 11
Aedes aegypti (mosquito) 190–1; *see also* mosquitoes
Aedes albopictus (tiger mosquito) 190–1; *see also* mosquitoes
aerosols 24, 36, 121, 209
aflatoxins 29–30
Ahmedabad HAP (India) 153–4
Ahmedabad Municipal Corporation (AMC) 153–4
air: 59, 60, 62–80, 87, 90–3, 95–9, 101, 104–7, 110, 113–4, 116, 119–22, 126, 138, 140, 152, 163, 178–9, 181–2, 199, 201–2; borne aerosols and dust 24, 150–1; borne diseases 6, 18–9, 33, 68–9, 76; chemistry 76–7, 144; cold 65, 80–2, 87, 92, 96, 103; conditioning 64; currents 68; density 103; hot 65, 92, 93, 181; pollution 5, 21, 29, 75–6, 95, 101, 130, 144, 171; pressure 70–1, 76–8, 81, 83, 87, 91–3, 95, 113, 116, 120, 132, 162, 181; quality 75, 95, 130, 144, 148, 150, 154, 171; saturated 67, 69–70, 93; temperature 20, 62–7, 69–70, 72–4, 77, 83, 93–4, 98, 102, 119, 136, 138, 140, 162, 179, 181; traffic 81, 162; turbulence 69; upper 69, 72, 162–3; warm 80–1, 93, 103
albedo effect 77, 99, 181
allergens 5, 21
Alps 93
American Meteorological Society 61, 75
anaemia 27–9
analysis 138, 140, 160, 162–5, 183, 185
Anansi and the chameleon 102
Antarctica 75, 96, 201–2; *see also* South Pole
anthrax 26
apparent temperatures 23, 64–5, 74
Arctic 75, 98; *see also* North Pole
Arica, Chile 156
Arrhenius, Svante 201
Asia-Pacific Economic Cooperation Climate Center (APCC) 190
asthma 18, 148
Atlantic Equatorial Mode 119
Atlantic Ocean 180, 203, 214
atmosphere 5, 60–2, 67, 69, 75–7, 100, 102, 114, 115, 121–2, 149, 160, 162, 183, 185, 200–1, 205, 208, 215
Atmospheric Circulation Reconstructions over the Earth (ACRE) 131
atmospheric composition 122
Australia 30, 180
avian influenza (H1N1) 1

balloons 162
Bangkok Principles 6
Bangladesh 22, 80, 203
barometric pressure *see* climate, and air pressure
Bayesian inference algorithms 160–1
Beaufort scale 74
Black Death *see* plague
blackflies 69
blocking 107, 152
Bodélé Depression, Chad 150
Botswana 11, 49, 118, 187

boundary conditions 184
boundary forcing 178, 182, 186
Brazil 47, 177, 190–91
bubonic plague *see* plague
Buenaventura, Colombia 156

California 5, 61, 98, 110, 181
Campylobacter 32
Canada 23, 82
carbon dioxide (CO_2): and climate adaptation 200–5; concentrations of 200–5; increases in atmospheric 21, 122–3; and plant growth 205–6, 208; and seawater 208–9; *see also* greenhouse gases (GHG)
Center for Disease Control 10
Central China flood disaster (1931) 21
Centre for Research on the Epidemiology of Disasters (CRED) 24
Centre for Reviews and Dissemination (CRD) 51
cereals *see* crops
Chicago heat wave (1995) 107
chikungunya virus 190
children 7–8, 27–8, 30, 32, 34–7, 37, 47, 137
China 21, 107
chlorofluorocarbons (CFCs) 75
cholera 1, 33, 61, 203; *see also* infectious diseases
Chronic Obstructive Pulmonary Disease (COPD) 148
climate: climatology 82–7; and crops 31–2; definition of 59–61; education and training 230–3; and health 42–56, 220, 224; impact monitoring groups 17; impact on agriculture 29–30; institutional arrangements 228–30; and livestock 26; modelling of 53–6, 183–4; and population vulnerability 37; public health action 219–20; seasonal and nutrition 32–5
climate adaptation information: accuracy of forecasts 213–15; and CO_2 concentrations 200–5; downscaling 209–210; and the global public health community 5–11; multi-annual/decadal prediction 211, 213; and nutrition 205–9; projections 209
climate data: availability of 140–4; modelled 138, 140; production/sharing of 126–30; types of 130–4, 136, 138, 140
climate-driven models 51–4, 138, 140, 213
Climate and Health Country Profile reports (WHO) 211–12
Climate Prediction Center/IRI joint ENSO Diagnostic Discussion 103
climate services 11–13, 17, 56, 151, 190–2, 220–4, 227, 229, 230, 232–3
Climate Services for Health (guide) 232
climate stations 130–2
climate variability: altitude 90–6; duration of weather patterns 107; effects of deforestation 102, 200; effects of land and sea 99–101; external causes of 120–3; latitude 96–9; seasons 108–113; spatial scale 102–5; temporal variance of 113–20
climatology 82–7
cloud monitoring 135
cloud-top temperatures 134, 136, 138
coastal flooding 22–3, 142, 208
cold: extremes 18, 23, 74, 76, 82, 107, 122, 148, 168, 171, 200, 212, 222; spells 104, 171; winds 168, 171
Colombia 117, 231–2
convective rainfall 66, 104, 106, 107, 110, 120, 134, 168, 180
cooling degree days 63
Coordinated Regional Downscaling Experiment (CORDEX) 214
Coriolis effect 78
coronavirus 1
counter-factual analyses 56
Coupled Model Intercomparison Project Phase 5 (CMIP5) 211–12, 214
crops 7, 26, 29–32, 46, 76, 144, 195, 208
Cyclone Eline 49
cyclones: 18, 33, 38, 74, 104, 222; and boundary forcing 186; extratropical 80–2, 104–5, 213; tropical 78–81, 87, 129, 142, 162, 168–71, 202–3; warnings of 148

data rescue 131
data sharing 45–6
decadal prediction *see* multi-annual prediction
Decade for Action of Nutrition 7
deforestation 102, 200
Delphi review 51
dengue fever 18, 20, 24, 25, 33, 177, 190
deterministic forecasts *see* forecasts
deterministic models 54
digitization software 131
disaster risk reduction (DRR) 6–7
disease modelling 54
District Health Information System II (DHIS2) 228
diurnal variability 105–7

DNA sequence 35
downscaling 48, 128, 186, 188, 209–10; *see also* Coordinated Regional Downscaling Experiment (CORDEX)
drones 138–9
dropsondes 162
drought 5, 6, 7, 9, 18, 19, 21, 23–4, 26, 29, 30, 31, 35–6, 51, 85–7, 102, 117, 119, 141, 143–4, 171, 180, 190, 192, 201, 207
dust-storms 148–51
dynamical modelling/prediction 70, 160, 182, 183, 185–6, 188, 209–10

early warning systems (EWS) 44, 47, 49, 53, 55–6
earthquakes 55
East Africa 85, 94, 110, 127, 207–8, 214
East Africa Paradox 207, 213
Ebola virus 1, 10
El Niño Southern Oscillation (ENSO): accuracy of prediction 45; areas beyond the Pacific 180; and climate change 200, 206; and climate data 132–3, 143; and climate variability/trends 111, 113, 115–19; and climatological period updating 83; and forecasts 182; and Global Seasonal Climate Update 193; and health 220; international responses to 224; and seasonal forecasts 193; and variability in tropical Pacific sea-surface temperatures 179–80; and western Pacific typhoons 79; and ZIKV epidemic 190–1
elderly people 7, 37, 42
emergency and disaster risk management programmes (DRM) 6
Emergency Events Database (EM-DAT) 24
Empirical/statistical prediction/modelling 182
Enhancing National Climate Services (ENACTS) 127–8, 138
ensemble forecasting 164–6
Epic of Gilgamesh 175
epidemics: cardiovascular disease/respiratory problems 150–1, 171; chikungunya/yellow fever viruses 190–1; climate/health outcomes 55–6; countering 155; dengue fever 177, 190; and El Niño and La Niña 117, 119; general 19–25; influenza 38, 226; introduction 1, 5–6; malaria 29, 44, 191; visceral leishmaniasis 141; Zika virus 16, 47–8, 206
equinoxes 109
Ethiopia 6, 94–5, 117
Europe 23, 48, 51, 107, 120–1, 144, 148, 152, 181
European Centre for Medium Range Weather Forecasts (ECMWF) 164
European heat waves (2003/2010) 107
extratropical cyclones 80–2, 213
extratropical oceans 181

fertilization effect 31–2, 201, 205–6, 208
filarial worms 19, 25, 72
fleas 48
floods 1, 18, 22–3, 32, 134, 142, 208
fog 67, 70, 93, 113
food production 1, 7–8
Forecast-based Financing (FbF) 22
forecasting, ensemble 164–6
forecasts: deterministic 154–8, 164, 167, 176, 187; ensemble 164–6; probabilistic 56, 152–5, 157–8, 161, 167, 176, 187, 194; seasonal 2, 37, 128, 149, 172, 176–9, 181–96; statistical 48–50, 161, 166; sub-seasonal 149, 151–2, 1 67, 178
fossil fuels 1, 5, 123, 200
France 82, 86
frontal rainfall *see* large-scale rainfall
fronts 80–1, 103–5, 168

G. morsitans morsitans see tsetse flies
Gambia 35
general circulation models (GCMs) 209–210, 214
geostationary satellites 133
glaciers 208
Global Atmospheric Watch 130
Global Climate Observing System (GCOS) 126
Global Consortium on Climate and Health Education (GCCHE) 232
Global Data Processing and Forecasting System (GDPFS) 128–9
Global Framework for Climate Services (GFCS) 11–13, 126, 143, 189, 192
Global Fund to fight AIDS, TB and Malaria (GFATM) 4
global health challenges 2
Global Health Security Agenda (GHSA) 9–10
Global Nutrition Report (2015) 8
Global Observing System (GOS) 126
Global Precipitation Climatology Centre (GPCC) 127–8
Global Producing Centres of Long-Range Forecasts (GPCs) 189–90
global public health community 5–11
global sea levels 208
Global Seasonal Climate Update 192–3

Global Telecommunication System (GTS) 128, 162
global warming 194–5, 201–2, 213; *see also* climate adaptation information; carbon dioxide (CO_2); greenhouse gases (GHG)
Glossina pallidipes see tsetse flies
Google Flu Trends (GFT) 226
Green Climate Fund (GCF) 13, 229
Green Revolution 31
greenhouse gases (GHG): atmospheric composition 122; basics 76; and climate-change signal 200; crop production/supply chains 7; increases in emissions of 5; introduction 1; projections 209; and solar activity 122; *see also* carbon dioxide (CO_2)
gridded station data 131–2
growing degree days (GDD) 63–4
Guillain-Barré syndrome 47
Gulf of Guinea 180

halons 75
Harmattan 150
health: and agriculture 3, 8, 46, 219; anaemia 27–9; and climate 42–56, 220, 224; and dust storms 150–1; and El Niño 220; global challenges 2; human monitoring/warning systems 148; human pharyngeal mucosa 34; kwashiorkor disease 30; Latin Aedes borne diseases 190; leptospirosis 26; low birthweight (LBW) 28–9; lymphatic filariasis 25; malnutrition 28; meningitis 25, 34, 140, 148, 150–1, 171; and non-communicable diseases (NCDs) 7; and onchocerciasis 25; and Onchocerciasis Control Programme (OCP) 72–3; pathogens 25, 33, 68; plague 1, 48; and public health action 229–33; and temporal variability of climate 214; typhoid 33; and vector-borne diseases 47–8; *vector-transmitted infections* 33; and yellow fever virus 190; and Zika virus (ZIKV) 1, 16, 47, 190–1, 206; *see also* infectious diseases; malaria
heat: action plans 153–4; and climate monitoring 143; early warning systems 152–3; heatwaves 23, 42, 56, 102, 148; indices 64, 171; latent 65; relative 84–7
Heat Early Warning Systems (HEWS) 148
heating degree days 63
Himalayas 93, 119–20, 181
hindcasts 189
Hippocrates 231
historical proxy datasets 133
Hong Kong 63
human health monitoring and warning systems 148
human pharyngeal mucosa 34
humidity: and altitude 95; and climate 67, 69–72; in coastal areas 100–1; and the maximum temperature 105
Hurricane Harvey (2017) 80, 202
Hurricane Irma (2017) 82
Hurricane Mitch (1998) 80
hurricanes: and 2017 season 202–3; and changes of wind patterns 203; description 78; measuring 74; and weather forecast utility 148, 161; *see also* tropical cyclones
hydro-meteorological disasters 21–4, 202–3

ice 181, 185, 208
Ice Age 122
index datasets 132
India 31, 153–4
Indian Institute of Public Health-Gandhinagar (IIPH-G) 153
Indian Meteorological Department (IMD) 154
Indian Meteorological Society (IMS) 154
Indian monsoons 182
Indian Ocean 35, 117, 180, 207
Indian Ocean Dipole 119
infectious diseases: and climate change 203, 205; climate impacts on 5–6; forecasting systems 160–1; influenza 33–4, 68, 160–1; models 53; nature of 24–7; and seasonal fluctuations 33–5; transmission dynamics of 143; *see also* health; malaria
information and communication technologies (ICT) 224, 226
initialization 163, 185
insect vectors 68–9
integration 163–4, 186
Intergovernmental Panel on Climate Change (IPCC) 126, 202, 207, 211
International Center for Diarrhoeal Disease Research (ICDDR) 203
International Cochrane Collaboration (ICC) 51
International Health Regulations (IHR) 4
International Research Institute for Climate and Society (IRI) 11, 117, 127, 143, 179, 190, 193–4, 206
International Telecommunications report (ITU) 227

internet *see* World Wide Web (WWW)
inversions 93–5
IPCC Assessment Reports 126, 143
iron deficiency *see* health, anaemia
iterated filtering-type approach 53

jet streams 77, 80, 107, 113, 163, 181
Joint Office on Climate and Health (WHO/WMO) 17

Keeling Curve 202
Kelvin scale 65
Kenya 53, 229
kwashiorkor disease 30

La Niña 83, 116–18, 132–3, 179–80, 182, 193, 203
La Paz, Bolivia 76
lake effect snow 101
Lake Victoria 148
Lancet Countdown 233
large-scale rainfall 66, 80, 104, 106, 107, 110, 168, 213
latent heat 65, 78–9, 81, 93, 106, 179
Lead-Centre for Multi-Model Ensembling 189
lead-times 149
leptospirosis 26
low birthweight (LBW) 28–9
low- and middle-income countries (LMICs) 7
lymphatic filariasis 25

McMichael, Tony 5
Madagascar 34
Madden-Julian Oscillation (MJO) 120
malaria: and climate change 203, 213; control programmes 94–5; early warning systems for 190; health priorities 2, 9; interactions between temperature and 49; and La Niña 117–19; modelling of 53; and mosquito bites 25; in pregnancy 29; and Roll Back Malaria 11, 207; seasonal 137; transmission season of 220; vector-transmitted infection 33, 36; and Web platforms 227–8; *see also* health; infectious diseases
Malaria Atlas Project (MAP) 25
Malaria Outlook Forum (MALOF) 191
Mali 68
malnutrition 28
Managing the Health Effects of Climate Change (Lancet Commission) 199
Markov Chain Monte Carlo model 53
Massively Open Online Courses (MOOCs) 226
mathematical models 53; *see also* dynamical models; empirical models
Mauna Loa Observatory, Hawaii 201–2
maximum likelihood approaches 53
Mectizan (ivermectin) 72
Meikirch Model of Health 19
meningitis 25, 34, 140, 148, 150–1, 171
Meningitis Environmental Risk Information Technologies initiative (MERIT) 150
Met Office (UK) 148, 209
Meteo Rwanda 127
MeteoAlarm 170
methane 122–3; *see also* carbon dioxide (CO_2)
microcephaly 47
Microsoft 139
microwave wavelengths 136
Middle East 64, 110, 120, 144, 150
migration 9, 24, 34, 37, 107
Millennium Development Goals (MDGs) 2–4, 8, 225–7, 229
Millennium Summit of the United Nations (2000) 3
millet *see* crops
model output statistics *see* MOS guidance
modelled climate data 138, 140; *see also* analysis; reanalysis
modelling: climate data 138, 140; deterministic 54; disease 54; dynamical 70, 182–3, 185–6; empirical 70, 182; empirical prediction 182; output statistics *see* MOS guidance
Mongolia 82
monsoons 71, 101, 182
Montreal Protocol (1987) 75
MOS guidance 166, 188
mosquitoes 25, 34, 42, 47, 69, 139
Mount Pinatubo volcano (1991) 121
Mount St. Helens volcano (1980) 121
multi-annual prediction 211, 213–14, 224
multi-model ensembles 186
municipal Heat Action Plan-India (HAP) 153
mycotoxins 30

National Adaptation Plans (NAPs) 5
National Center for Atmospheric Research (NCAR) 48, 209
National Disaster Management Authority (NDMA) 154
National Health Service (UK) 51, 148

National Institute for Health and Clinical Excellence (NICE) 51
National Meteorological and Hydrological Services (NMHSs) 126, 128, 143, 147, 162, 170–1, 189, 192, 224
National Oceanic and Atmospheric Administration (NOAA) 143
Natural Resources Defense Council (NRDC) 153–4
near-infrared wavelengths 135
New Delhi 95
New York 84–5, 95
Niger 36, 72, 151
Niño3.4 index 133
NOAA monthly Global Climate Reports 143
nodal officers (India) 154
non-communicable diseases (NCDs) 7
Normalized Difference Vegetation Index (NDVI) 135
North American Multi-Model Ensemble (NMME) 191
North Atlantic 36, 74, 78, 142, 169, 202–3
North Atlantic Oscillation (NAO) 120
North Pole 96–9, 109–110, 133; *see also* Arctic
Numerical Weather Prediction (NWP) model 163–6, 168–9, 183, 186
nutrition: and climate adaption 205–9; climate impacts on 7; and good health 27, 29

Oceanic Niño Index (ONI) 133
Onchocerca volvulus (filarial worm) 72–3
onchocerciasis 25
Onchocerciasis Control Programme (OCP) 72–3
open data policies *see* data sharing
orographic rainfall 66, 96, 106
ozone 75, 77, 91, 122

Pacific Ocean 115–17, 119, 132, 179–80, 184, 207
parameterizations 164, 209
Paris Agreement *see* United Nations Framework Convention on Climate Change (UNFCCC)
pastoralism 26
pathogens 1, 24–6, 30, 25, 33–4, 42, 63–4, 68, 71, 74, 154, 192, 195
Pearson Type-III distribution fit 86
Philippines 34
photosynthesis 75
plague 1, 48
pneumonic plague *see* plague
poison ivy photosynthesis 208
polar lows 81–2
polar-orbiting satellites 133
Pontificia Universidad Javeriana 231
post-processing 160, 166, 168, 183, 186–9, 210
precipitation *see* rainfall
PRECIS model 209
predictands 53, 55, 149, 187, 195
predictions: *see* forecasts
predictors 210, 220, 221
President's Malaria Initiative (PMI) 2
Program for Monitoring Emerging Diseases (ProMED) 225
projections: *see* multi-annual prediction
protein-energy malnutrition (PEM) 28
Public Health Foundation of India (PHFI) 153

radar 162
radiosondes 162
rainfall: and altitude 91, 96; anomalies 85; changes in evaporation 179; for defining the seasons 109–110; and crops; and deforestation 102; diurnal variability in 106; and El Niño 118; and extratropical cyclones 80–1; and flooding 18, 80, 81, 141–2, 155, 192, 209; forecasts 148, 156, 195; gauges 67; general 49–50; induced flooding in India 155; and the Inter-Tropical Convergence Zone (ITCZ) 110; large-scale 66; and latitude 96–7; and malaria 207; measurement of 84; and modelling 213; and Mount Pinatubo eruption 121; and pathogens 192, 195; rainstorms 107; and reanalyses 140; in Rwanda 127; in the Sahel 137; satellite monitoring of 134, 136, 138; and sea-surface temperatures 100–1, 119; and seasonal temperature forecasts 193–4; spatial scales of 25, 66, 104–5, 166; and temperature 112, 168, 214–15; and tropical cyclones 80; variations in 206; *see also* climate; convective rainfall; orographic rainfall; vector-borne diseases
Randomized Controlled Trials (RCTs) 50–1
rats 48
rawinsondes *see* radiosondes
reanalyses 73, 90, 91, 97, 98, 100, 103, 106, 111, 112, 127, 140, 220
Red Cross Red Crescent Climate Centre (RCRC) 22
Regional Climate Centres 143, 191–2

regional climate models (RCM) 209, 214
Regional Climate Outlook Forums (RCOFs) 83, 191–2, 195
Regional Specialized Meteorological Centres 171
remote sensors 135
respiratory syncytial virus (RSV) 34
rice *see* crops
Roll Back Malaria 11, 207
Rwanda 127–8

Sahel: climate's impact on social and economic development in 11; drought in 35–6, 180; dust storms in 150–1; dusty dry season 171; epidemics in 25, 34; monthly rainfall estimates 137; and the plague 48; rain in 137, 215
Salmonella 32
Sand and Dust Storm initiative (WMO) 150
satellite data 133–4, 138, 162, 220
sea levels 208
sea-ice 185
seasonal climates 176–89
seasonal forecasts 2, 37, 120, 128, 149, 151, 153, 158, 165, 172, 176–8, 181–96, 209–11, 222
seasonal malaria chemoprevention (SMC) 137
seasonality 32–5
sea-surface temperatures 100–1, 209
2nd International Conference on Nutrition (ICN2) 7
Semantic Web *see* Web 3.0
Sendai Framework for Disaster Risk Reduction (2015–2030) 6, 13
Severe Weather Information Centre (WMO) 170
Shanghai Food and Drug Supervision Administration 148
Shanghai Health Bureau 148
Shanghai Meteorological Bureau 148
Shanghai Municipal Center for Disease Control and Prevention 148
Sierra Leone 22, 72
Sierra Leone mudslide (2017) 22
Simulium damnosum s.l. 72
Small Island Developing States (SIDS) 208
smog (New York City, 1966) 95
snow 101, 181, 185
Socrates 147
Soil: and deforestation 102; evaporation from 77, 102, 181; and heat-waves 102, 107, 120; moisture 23, 77, 102, 107, 114, 120, 144, 152, 181, 185; properties 135, 185; rich 94; type 42
solar radiation 74–5
solar variability 121–2
solstices 108
sorghum *see* crops
South Asia 152, 181, 229
South Pole 96–9, 110, 133
Southern Africa 191
Southern Oscillation Index (SOI) 116, 133
Spot the Ball puzzles 177–8
standard operating procedures (SOPS) 224
Standardized Precipitation Index (SPI) 86–7
statistical models 50–2; *see also* empirical models
stochastic models 54
storms: movement and evolution of 163; prediction of 167; tropical 168–70; warnings 148; *see also* cyclones
Streptococcus pneumoniae bacteria 151
sub-seasonal: forecasts 45, 149, 151–3, 167, 178, 185, 190, 196, 222, 224; variability 43, 114, 214
Super Typhoon Tip (1979) 81–2
supply chains 7
Sustainable Development Goals (SDGs) 3–4, 8, 10–11, 226–7, 229
sweating 67
synoptic circulation systems 23
Syria 9

Tanzania 207
target periods 149
Technical Expert Group on Preventive Chemotherapy (WHO) 137
technology advances 224, 227–8
teleconnections 180
temperature: air 20, 62–7, 69–70, 72–4, 77, 83, 93–4, 98, 102, 119, 136, 138, 140, 162, 179, 181; air surface estimates from reanalyses 140; airborne diseases 6, 18–9, 33, 68–9, 76; and altitude 90–5; apparent 64–5; between night and day 105–6; cloud-top 134, 136, 138; and cold extremes 168; disease transmission 47, 62, 160–1, 191, 203; predictions of 168, 213–14; satellite monitoring of 138; sea-surface anomalies 114, 119, 179, 195; spatial scales of 103–4; thresholds 23, 56, 93, 154; variability of 110–113; and vector-borne diseases 47–8
tercile forecasts 187–8, 192
terciles 85, 187
Texas floods (2007) 107

thermal infrared wavelengths 135–6
tornadoes 74, 78, 82, 154
torrid zone *see* tropics
Touregs 64
Trade Winds 71, 116
Tropical Atmosphere Ocean/Triangle Trans Ocean Buoy Network array (TRITON) 184–5
tropical cyclones 78–81, 168–70, 202–3
tsetse flies 203–4
tuberculosis (TB) 2, 4
Tyndall, John 201
typhoid 33; *see also* health; infectious diseases
Typhoon Koppu (2009) 80
typhoons 78–9, 148, 168–70, 203

Uganda 48
Ukraine 30
ultraviolet radiation 74, 75, 91, 130
United Kingdom: and 2016/17 winter 86; and cooling/heating degree days 63; Great Smog (London, 1952) 95; heat alerts in 56
United Nations Children's Fund (UNICEF) 28
United Nations Educational Scientific and Cultural Organization (UNESCO) 203–4
United Nations Framework Convention on Climate Change (UNFCCC) 5, 8–9, 13, 199
United Nations International Strategy for Disaster Reduction 22
United States: and 2017 hurricane season 202; backing away from the Paris Accord 5; and cooling/heating degree days 63; and the 'Dust Bowl' (1930s) 181; and extreme smog 95; and heat waves 23; and infants conceived in May 34; leading the GHSA 10; and North Atlantic Oscillation (NAO) 120; and predictability of heat waves 152; and winds 103–4
United States Agency for International Development (USAID) 2
United States Centers for Disease Control and Prevention (CDC) 48
United States Drought Monitor (USDM) 87
Universal Health Coverage (UHC) 8–9
unmanned aerial vehicles (UAVs) *see* drones
upper-air soundings 162
urban heat islands 101–2
urushiol 208

vector: borne diseases 1, 5, 6, 18–20, 22, 27, 47–8, 155, 172; breeding sites 36, 49, 72; climate sensitivity of 25, 26, 64, 67–9; control 44, 72–3, 139; development states 63; examples of 24, 36, 47, 48, 74, 190–1, 203; processes of interaction 160; spread of 1, 71, 190; transmitted infections 33
vegetation monitoring 135
volcanoes 76, 114, 121, 129, 159, 171, 178, 193

water: and agriculture 42; and coastal flooding 208–9; converting to vapour 65, 67, 93, 100–1, 122; and drowning 47
water-borne diseases 6, 18, 20, 26, 33, 142
water vapour 65–6, 67, 70, 78, 80, 93, 95, 96, 100–1, 122, 179, 201
wavelengths, *see* near-infrared; thermal infrared 135–6; ultraviolet 74–5, 91, 130
weather: aggregating data 83–4; blocking patterns 107; and climate-sensitive disease models 54–5; definition of 60; expectations of 82–3; sub-seasonal forecasts 151–2
weather forecasts: accuracy of 166–70; and climate forecasts 149–50; formats 158; and the health community 148, 151, 153–4; in India 155; and influenza 160–1; one-tiered forecast systems 184; prediction problems 159–66; two-tiered forecast systems 183–4; verification of 155–7
Web 3.0 226
Weighted Anomaly Standardized Precipitation Index (WASP) 87
winds 67, 71–82, 84, 93, 99, 101, 103–7, 110, 111, 113, 116, 119: and altitude 95; and climate variability 106–7; and dispersion 68, 75, 95; effect of time of day 107; in the extratropics 103–4; extremes 59, 78, 82; wind chill factors 37, 64–5; wind-related injuries 18
WMO Annual Statement on Status of the Global Climate 143
WMO Resolutions 129
women 3, 7, 27, 29, 37

World Bank 8, 22
World Climate Conference III (WMO) 11–12, 129, 143, 170
World Cup (2014) 177
World Health Assembly 5, 199
World Health Organization (WHO) 4, 17, 24–6, 137, 150, 199
World Health Organization Collaborating Center 190
World Meteorological Organization (WMO) 17, 87, 126, 128–30, 143, 149, 162, 170, 189, 192–3
World Weather Watch 126, 171
World Wide Web (WWW) 225WRF model 209

Xenopsylla cheopis see fleas

Y. pestis bacteria 48
yellow fever 20, 190

Zambezi Valley 203
Zika virus (ZIKV) 1, 10, 16, 18–20, 27, 47, 190–1, 206
Zimbabwe 203
zoonosis 26